Suckler Herd Health and
Productivity Management

Suckler Herd Health and Productivity Management

Keith Cutler BSc BVSc DipECBHM MRCVS

 THE CROWOOD PRESS

CONTENTS

PREFACE

This book is written based on science but influenced by the 'art' and experience gained from over thirty years working as a veterinary surgeon with farmers and their cattle in the central south and south-west of England. In such a position, most vets have a heavy dairy bias, and I, too, have spent a significant part of my professional life working with dairy cattle. I have, however, a passion for beef cattle, and particularly beef suckler cows, which influenced my early career. My mornings would be spent on dairy farms. In the afternoons, looking to fill my time with something more productive than sitting in the office drinking coffee and waiting for an emergency call to come in, I would drive miles to one of my friends and clients running large suckler herds, numbering several hundred and up to a thousand breeding cows, on Salisbury Plain, lie through my teeth saying that I was just passing and ask how the cows were and whether we could have a look at them (free of charge, of course!).

Word gradually got round that there was this unusual vet in the area with an interest in suckler herd management, health and productivity, and work gradually built up until my suckler-cow work expanded sufficiently to displace the morning dairy-cow work.

There is much that the veterinary profession, armed with science and experience, can offer beef suckler farmers to assist with herd health management aimed at improving productivity performance; the veterinary work in beef suckler herds includes so much more than just TB testing and difficult calvings! An increased veterinary involvement should be welcomed by suckler farmers and viewed as an asset of value rather than simply a cost. The veterinary profession, for its part, perhaps needs to reconsider its attitude to suckler herd management, and stop viewing suckler cows as the poor relations of dairy cows!

INTRODUCTION

A suckler herd, or what in America is called a cow-calf operation, is one in which the primary output, unlike dairy herds where it is milk, is weaned calves usually destined for fattening for slaughter for human consumption, but which may be destined for retention or sale as breeding animals. Additional outputs also include 'barren' cows deemed to have reached the end of their economically productive life, again usually for fattening or immediate slaughter for human consumption, and the sensitive management of environmental or ecologically important ecosystems.

Herd size and preferred breeds vary widely dependent on owner preference and desired outcome. Most commercial suckler herds may number anything up to many hundreds

A Chillingham bull grazing traditional woodland pasture.

A Limousin cow and calf on parkland grazing.

of head, and will be made up of cross-bred cows, often originating from the dairy herd, to capitalise on the benefits of hybrid vigour; pedigree herds, often with the aim of producing terminal sires with superior genetic potential for use in commercial herds, are smaller, and 'niche' herds of smaller or rare breed animals are not at all uncommon.

Irrespective of the composition of the herd, output should target a calf per cow each year that is delivered easily, suckles vigorously, grows quickly, and is successfully weaned at the desired weight and at the desired time. In reality, however, this is probably rarely achieved. Breeding strategy will usually involve natural service, although the benefit that can be realised from the genetic potential available when using artificial insemination, either to observed or induced oestrus, should not be discounted without consideration. In most suckler herds, however, the widespread use of natural service means that far more attention must be paid to male fertility than is the case in most dairy herds, where vast amounts of time and effort are concentrated on managing female fertility.

Health considerations in the suckler herd also need to be prioritised differently to those in the dairy herd. The multifactorial issues, including mastitis and lameness that are major challenges to the dairy industry, are of lesser significance, although perhaps no less important, in the suckler herd, whereas many of the single-agent infectious diseases, particularly those that impact longevity and fecundity, assume arguably greater importance in the suckler herd.

CHAPTER 1

THE ETHICS OF SUCKLER PRODUCTION

Before discussing 'how', time should perhaps be spent considering 'why', and the ethics of suckler production. There are those who consider the 'exploitation' of any animal for the benefit of humankind to be abhorrent; there are others who consider, for whatever reason, that eating meat is wrong; and there are those who believe, in these days of increased concern about climate change and global warming, that we should not be farming ruminants. Others, perhaps the majority, disagree in every respect.

The use, and therefore probably misuse, of animals to suit our purposes has been the case for millennia. This does not make it right, but in many situations, including in animal agriculture and in the production of food for human consumption, the majority view is that the keeping, breeding and eventual slaughter of animals to provide food for people is acceptable – assuming that animal welfare is given the highest possible priority, and that every effort is taken to prevent suffering.

It is, of course, true that animal protein is not essential in the human diet to maintain health and vigour: plant-based proteins offer an alternative, as do other sources – insects, for example, or even laboratory-cultured protein, although these currently may not be economically viable alternatives. We are faced, however, with a growing global human population that it is going to be challenging to feed adequately. So why grow crops, which could be used to feed people, to feed to animals, including cattle, to then slaughter them for human consumption? Why not just eat the crops our-selves? This argument becomes problematic when the land area available for growing crops is not enough to produce sufficient plant-based protein to feed the growing global human population.

We could, of course, clear more land for agriculture, but at what expense to our environment? An alternative is to use land unsuitable for growing crops to produce food, and currently the only species able to do this, albeit perhaps somewhat inefficiently and at a relatively low level (but some, surely, is better than none!), are ruminants, essentially cattle, sheep and goats; this is thanks to their highly specialised digestive tracts in which fibre of a quality too poor to sustain humans and other monogastric species can be fermented within the rumen to sustain and grow the animal. (Using ruminants to graze such areas can also be advantageous in the maintenance of environmentally valuable ecosystems and biodiversity.)

But are ruminants not responsible, thanks to this fermentation within their rumen, for the production of large quantities of greenhouse gases being shed into our atmosphere, and therefore for global warming and the climate catastrophe we face? It is true that ruminant digestive processes do result in the production of greenhouse gases, with the amount being

Angus suckler cows and calves grazing land unsuitable for crop production in the borders of Scotland.

produced depending on what the ruminants are being fed and what they are eating. However, ruminants are not new to the planet: vast, although perhaps somewhat depleted, herds of buffalo, wildebeest and antelope still graze across the African plains; it is not so long ago that similarly vast herds of bison ran wild across the grasslands of America; and shortly before that, in evolutionary terms, Aurochs (large native wild cattle) roamed Europe. These were, and are, all ruminants producing greenhouse gases as part of their existence. Perhaps global warming and climate change are not the fault of ruminants!

In addition there is also an increasing worldwide demand for ruminant protein, particularly high quality and 'pasture-fed' ruminant protein, with some cuts of meat from the best animals commanding a high price (consider a juicy rib-eye steak, a tender fillet or a cut of Wagyu beef).

Suckler farming, at least for the foreseeable future, would seem to be here to stay. It is our responsibility, therefore, to make it as welfare friendly, sustainable, and efficient as possible.

CHAPTER 2

NUTRITION

Ruminants have evolved over very many hundreds of thousands of years to survive, even thrive, on a relatively low-quality fibre-based diet that is fermented in the rumen and reticulum, the 'first' and 'second' stomachs, by a numerous and diverse flora and fauna. It may, of course, be possible that productivity can be improved by managing pastures and manipulating dietary constituents to increase the plane of nutrition, as is seen particularly in intensive beef fattening and high-yielding dairy enterprises. This does

not, however, come without cost, with the occurrence of metabolic diseases including ruminal acidosis, displaced abomasums and liver abscessation, to name but a few. The key is that in order to maintain a healthy cow we need to maintain a healthy rumen, and this relies on maintaining a healthy rumen flora and fauna.

The rumen flora and fauna, the 'bugs', comprise many different bacteria, yeasts and protozoal species: these require a stable environment maintained around a normal

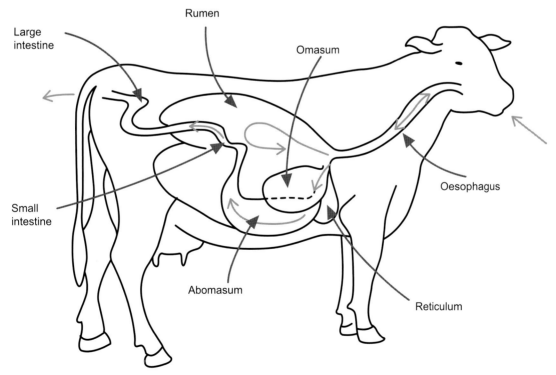

The bovine digestive system.

pH if they are to function optimally. This requires a consistent intake of forage to add to the mat of fibre that sits floating on the rumen liquor. Fibre from this mat is periodically regurgitated and masticated – 'chewing the cud' – to break it down mechanically before it is re-swallowed. Any sudden alterations to dietary intake, feeding discrete amounts of concentrate (essentially rapidly fermentable starch) for example, as may occur at milking time for dairy animals, will alter this constant rumen environment, usually reducing the pH, and have a deleterious effect on the bugs. Evolution has, however, predicted man's need to meddle, and has created a system where saliva, which is produced in copious quantities during rumination, contains bicarbonate to 'buffer' rumen pH and maintain the optimal constant environment.

As well as feeding the cow, to maintain health we also need to feed the rumen bugs. Just as it is possible to feed too much high-quality material it is also possible to feed diets of such low quality that the bugs do not have access to the nutrients they need to survive. Rumen function then literally 'grinds to a halt'. There may appear to be sufficient food for the cows to eat, and they may be eating it, but if it is of such low quality that the rumen bugs stop working, the fibre will not be broken down: the rumen will be full and getting fuller, becoming increasingly impacted and dysfunctional, because the cows will still be hungry, but they will lose weight and productivity will crash. While the answer in such a situation may seem simple (improve the quality of the diet), achieving a resolution is often anything but!

Water too, of course, is an essential part of the diet, and access to an adequate supply of potable water is vital to maintain health.

Nutrition of the calf is, of course, completely different. Although calves will start nibbling at forage at an early age, in the suckler herd situation the entirety of their nutrition during the earliest part of their life is totally dependent on milk from their mother. This continues to

Access to an adequate supply of potable water is just as important in the maintenance of health and the promotion of productivity as is easy access to a sufficient supply of a well-balanced diet (although when that access is from a naturally flowing water course don't forget the possible biosecurity risk!).

provide a significant part of their diet for many weeks and even months. However, milk intake is dependent on its production by the dam, which in turn depends on, amongst other things, a healthy rumen and maternal nutrition – and this must not be overlooked!

ENERGY

Energy supply, both directly to the cow but most importantly to the rumen bugs, is provided by the plant-based carbohydrates, including sugars, starches, hemicelluloses (cell wall polysaccharides) and cellulose, available in the diet. Although some of the simpler sugars present and produced during digestion can be absorbed and metabolised by the cow, the vast majority of the carbohydrates present in the diet are digested and fermented by the bugs producing gas, both carbon dioxide and methane, which must be voided by the cow to the atmosphere by eructating (burping) if a potentially fatal bloat is to be avoided. There is also the action of volatile fatty acids, primarily acetic, propionic and butyric acids, which are absorbed through the mucosa of the multitude of papillae covering the rumen wall, which vastly increase its area and therefore absorptive capacity, and so provide the major energy source of the cow.

Once absorbed, acetic and propionic acids, unchanged, and butyric acid converted to beta hydroxy-butyrate, are transported to the liver in the hepatic portal vein. Within the liver the propionic acid is converted into glucose, which may be stored as glycogen or metabolised in the various tissues and organs of the body to drive the TCA cycle (tricarboxylic acid cycle or Krebs cycle): this converts adenosine diphosphate (ADP) to adenosine triphosphate (ATP), producing

Suckling provides almost the entirety of a young calf's nutrition, and the importance of an adequate, early intake of good quality colostrum to give a good start in life cannot be overstated.

water as a by-product, to provide energy to the cells making up these tissues and organs. (Where oxygen supply is limited, the less efficient 'glycolytic' pathway can also be employed to provide ATP.) Acetic acid and beta hydroxy-butyrate are used unchanged as energy sources.

Beta hydroxy-butyrate is also released when, in times of deficient dietary energy intake, body-fat reserves are mobilised to make up the shortfall, and it is often assessed to provide information about the nutritional status of the cow; an elevated circulating beta hydroxy-butyrate level indicates excessive fat mobilisation, predisposing to fatty liver and other metabolic disease, and an inadequate dietary energy supply – although in cases of starvation where all the body-fat reserves have been depleted, the circulating beta hydroxy-butyrate level will return to normal!

PROTEIN

Protein metabolism, if anything, is arguably more complex than energy metabolism! Dietary proteins need to be broken down to their individual amino acid constituents before they can be absorbed through the

Rumen fermentation and Volatile Fatty Acid synthesis

Rumen fermentation synthesising volatile fatty acids (blue) and the greenhouse gas methane (red) from dietary constituents (green).

intestinal wall to be used by the cow as the building blocks for new proteins. Some amino acids, the non-essential amino acids, can also be synthesised within the body, but this is not universally the case; some amino acids, including arginine, histidine, isoleucine, leucine, lysine, methionine, phenylalanine, threonine, tryptophan and valine, cannot be synthesised by the cow, or only synthesised in limited quantity, and so are considered essential dietary components.

Much, but not all, of the protein supplied in the cow's diet (rumen-degradable protein, or RDP) will be broken down by the rumen bugs. Some of the amino acids produced will then be utilised by those bugs to produce microbial protein during their usual metabolism and reproduction. (Some will also pass into the abomasum and small intestine to provide an additional nutrient source to the cow.) In addition, many of the rumen bugs can use non-protein nitrogen sources, often supplied to cattle in the form of ammonia-treated straw or sometimes urea prills, to synthesise microbial proteins that, as the bugs die, will pass into the abomasum and small intestine where they, too, will be digested by the cow to provide the bulk of the cow's protein nutrition.

However, some of the protein in the cow's diet (dietary undegradable protein (DUP), or by-pass protein) will pass through the rumen without being broken down. This is then digested in the abomasum and small intestine as it would be in monogastric

Glycolysis and the Tricarboxylic acid (Krebs) cycle

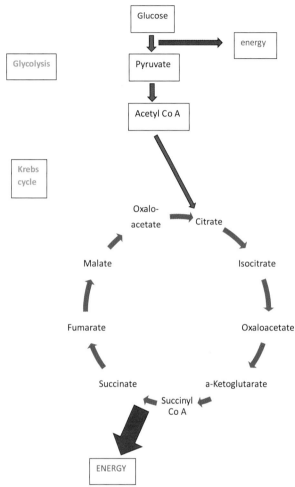

The glycolytic pathway and the tricarboxylic acid cycle – the power-house of energy provision for cellular aerobic respiration.

(non-ruminant) animals to provide additional amino acids.

As a last little twist, when amino acids are supplied to the cow in excess, they can be metabolised, mainly in the liver but also by the kidneys, to supply energy, producing ammonia as a by-product. Ammonia is, of course, toxic, and so if it is not used to produce more amino acids (the minority of the ammonia produced), it needs to be removed from the body. This is achieved by converting it into urea in the liver (it should be rapidly becoming clear what an important organ the liver is, and how important it is that it remains healthy!), which is then excreted via the kidney and voided in the urine.

FATS AND OILS

Fats and oils are, of course, a natural component of forages and can provide a very rich energy source. Additional fats and oils can be incorporated into cattle rations, usually those formulated for dairy cows where there is a high demand for energy to drive milk production. This can, however, come at a price, both literally and

Urea production

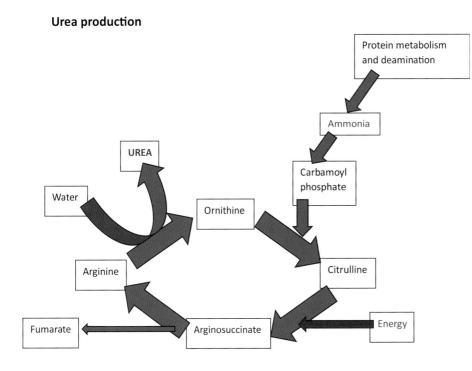

The production of urea, which can be safely excreted from the body via the kidneys by the detoxification of ammonia produced as a by-product of protein metabolism.

Mineral interactions

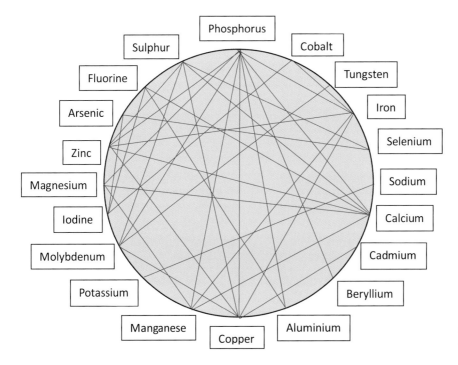

Mineral interactions are many and complex, making mineral nutrition a difficult area of the bovine diet.

metaphorically: fats as an energy source for ruminants are expensive (although perhaps not as expensive as purchased protein sources), and the inclusion of an excess amount in ruminant diets will compromise the efficiency of the rumen bugs and the functioning of the rumen.

VITAMIN AND MINERAL NUTRITION

While the various vitamins and minerals that are necessary for the maintenance of health and productivity are essential dietary components, and are often the primary consideration of many farmers and their advisers (if there is one thing that mineral salesmen are very, very good at it is their job!), they are only required in tiny amounts and over-supplementation can often cause as many or more problems than it resolves. The situation is often further complicated by multiple interactions and the difficulty of accurately establishing status in the presence of homeostatic mechanisms that continuously act to optimise circulating levels irrespective of the level of supply.

When supplementation is required, this can also present challenges: for example, what should be the route of supplementation, and what form should the supplement take? The easiest option may be to allow the cattle to help themselves from either free-access minerals or lick blocks (beware the amount of molassed minerals that may be consumed, and therefore the expense, because they taste nice!) but intakes will vary enormously. Injectable preparations or boluses will provide a more even and reliable level of supplementation to every animal but can, of course, be difficult and time-consuming to administer.

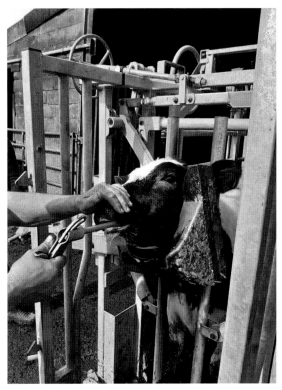

Bolusing cattle provides a more certain and reliable means of supplementing mineral status than relying on free-access preparations or lick blocks.

MACRO-MINERALS

Calcium
Calcium presents a conundrum. It is required for nerves and muscles to function efficiently and as a major component of bone to ensure skeletal integrity. Given the amount of calcium present in skeletal tissue, deficiency within the body is almost impossible, and yet, due to the time taken for enzyme systems to activate its release from skeletal tissue at calving when the demand from the calf and, more particularly, the mammary gland is maximal, a relative deficiency of circulating calcium, or hypocalcaemia, can result in muscular weakness, and in the most severe cases, a flaccid paralysis and recumbency, often termed 'milk fever'.

Furthermore, just as the skeletal muscles become weak in cases of milk fever, so too do the uterine muscles, so milk fever before the calf has been delivered may result in a prolonged birth. Furthermore recumbency after the calf has been delivered can be complicated by a prolapse of the uterus and inactivity of the smooth muscles of the gastro-intestinal wall, so bloat and constipation may also be seen. Cardiac muscle function may also be compromised, with death occurring in the most severe cases.

Although cases of milk fever are more common in dairy cattle than in suckler cows because of the vastly greater amount of milk they produce, suckler cows, particularly older, dairy-cross animals, can also be affected and may require rapid action. In the most severe cases, with the affected animal 'flat out' in lateral recumbency, this may require the immediate infusion of either a 20 per cent or a 40 per cent solution of calcium borogluconate intravenously (i/v). The advantage of a 40 per cent solution is that it contains twice as much calcium as a 20 per cent solution and so will correct the circulating calcium deficit faster – but beware, giving too much calcium too quickly may also be fatal, so for this reason it may be considered safer to use a 20 per cent solution for i/v administration and to use two bottles if twice as much calcium is required.

In less severe cases, where the cow is wobbly or if she is unable to stand, if she can maintain dorsal recumbency tradition has often been to give 'a bottle under the skin'. Here, too, 20 per cent solutions of calcium borogluconate are preferable to 40 per cent solutions because the greater osmotic pressure of the more concentrated

A recumbent cow shortly after calving showing the typical signs of milk fever: a flaccid paralysis with the head turned towards the flank. Urgent treatment with calcium is indicated.

solution requires dilution by body fluid before it can be absorbed into the circulation and so it is slower to act than a weaker solution. However, consideration should be given to the discomfort caused by the administration of large volume, subcutaneous injections.

The ideal solution may be to treat recumbent animals with an i/v infusion to rapidly address the situation, but to administer one of the liquids or boluses designed for oral administration to provide a source of calcium that is rapidly absorbed from the gut to provide a longer-term depot supply of the mineral that will persist until homeostatic mechanisms are up-regulated to release the required amount of calcium from the bones. Cows that are wobbly but still able to stand may only require oral dosing to prevent the situation deteriorating.

Magnesium

Unlike the situation with calcium, magnesium, which is also essential to maintain efficient nervous function, is only stored within the body in very limited amounts and so a constant dietary supply is required. Animals are at risk if intakes fall, or if gut transit time is decreased, giving less time for the magnesium that is present in the diet to be absorbed. Such a situation is often seen in the spring, particularly when cows are grazing heavily fertilised young rye-grass

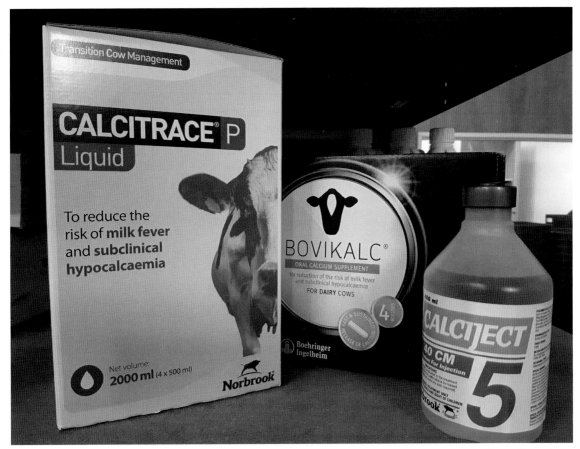

A wide range of milk-fever treatment options exists, including injectable solutions, oral liquids and boluses – but prevention, as they say, is undoubtedly preferable over cure!

A cow down due to hypomagnesaemia ('staggers') showing evidence of convulsions. Urgent treatment is required in such cases, but even when administered without delay, the outcome is frequently disappointing.

pastures when warm, wet weather promotes the rapid growth of lush grass that has both a low magnesium content and causes the animals grazing it to scour.

Animals affected by hypomagnesaemia will at first become nervous and twitchy. As the situation progresses they will start to become uncoordinated and will stagger around – hence the term 'grass staggers' – before becoming recumbent and fitting, with the limbs becoming rigid and thrashing about, and the neck and head often arched backwards over the shoulders. Such situations are medical emergencies: magnesium supplementation is urgently required, but i/v administration of solutions of magnesium sulphate is frequently fatal unless carried out with the greatest of care.

As usual, prevention is better than cure, and problems can often be avoided by increasing oral intake during high-risk periods by adding magnesium chloride flakes to the cows' drinking water (although these make it taste bitter and unpalatable so the animals will seek alternative water sources to drink from if these are available); providing 'high magnesium' lick blocks (which can be made more palatable, thereby increasing intake, if molasses is added); or bolusing the cows with magnesium 'bullets'.

Phosphorus

Phosphorus balance and metabolism is inextricably linked with calcium balance and metabolism, although it is not under such tight homeostatic control, and circulating levels can vary widely - including in the same animal in samples collected at the same time but from different sites; because of the massive excretion of phosphorus in the saliva of ruminants (the supply of phosphorus to the rumen in the saliva often exceeds dietary supply), samples collected by jugular venepuncture will contain lower levels than samples obtained from the tail - unless, of course the animal, for whatever reason, is not eating, ruminating or salivating!.

As with calcium, the role phosphorus plays within the body encompasses very much more than just skeletal integrity: it is essential in many vital roles, including energy metabolism. Deficiency, which is rarely reported in the UK, especially where animals graze or are fed forage harvested from fertilised pastures, can be a cause of inappetence and poor performance, and in chronic situations can result in affected livestock displaying 'pica' - eating abnormal materials (although this is not limited to deficiencies of this mineral alone). When present in excess (or when calcium and phosphorus supply is not correctly balanced, which not uncommonly occurs in young, rapidly growing animals) there may be a predisposition to the formation of urinary calculi, which, whilst not a problem whilst they remain within the bladder, can cause an acute and life-threatening situation if they enter and block the urethra, particularly of male calves.

Sodium, Chlorine and Potassium

Although perhaps not often considered in detail in the feeding of suckler cows and beef fattening animals, perhaps because natural deficiencies, particularly of chlorine and potassium, are rare, an adequate status of all of these minerals is vital to ensure good health and optimal productivity. On the other hand, an excess, particularly of potassium, may predispose to other problems including hypocalcaemia or milk fever around the time of calving, and hypomagnesaemia or grass staggers when lush pastures are grazed during periods of stress, including sudden inclement weather.

Sodium in particular has been referred to by Michell in 1985 in *The Veterinary Record* as providing the 'osmotic skeleton' of the body on which all else hangs. A shortage of sodium may be manifested as a roughened coat (how often are salt licks recommended, whether reasonably or not, for animals in the spring that fail to shed their winter coat?), a reduced appetite (adding salt to feed is an easy way of improving palatability), weight loss and a reduced milk yield.

Supplementation by providing access to manufactured salt blocks or lumps of natural rock salt for the animals to lick is, fortunately, simple and cheap, will promote salivation to help buffer the rumen to ensure the health of the bugs, and can be a useful, if somewhat unpredictable means of ensuring an intake of other essential minerals and vitamins.

Rock salt is a useful source of minerals, particularly iodine, as well as sodium and chloride, and will stimulate salivation to buffer rumen pH.

MICRO-MINERALS AND VITAMINS

Cobalt

Cobalt is required in ruminants for the synthesis of vitamin B12 by the microbes within the rumen. (Non-ruminant species lack this ability and require dietary vitamin B12 rather than cobalt to maintain health. The consequences of inadequate cobalt availability to ruminants, however, can be addressed by the administration of vitamin B12).

Assuming an adequate dietary supply of cobalt, any vitamin B12 deficiency and the consequent listlessness, inappetence, anaemia, poor performance and eventual emaciation and death, as would also be expected in cases of generalised malnutrition and starvation, are rare occurrences in cattle. There has, however, been a suggestion that a suboptimal cobalt status might be responsible for poor reproductive performance in beef cattle, although the assessment of cobalt status, despite the availability of a blood test to assess vitamin B12 status, is challenging.

Copper

Copper is complicated! It acts as a co-enzyme in many enzyme systems, is necessary for the production of red blood corpuscles and the normal functioning of the immune system, is essential in the development of myelin, which surrounds neurons within the central nervous system (deficiency in this situation is particularly well known and documented as causing swayback in lambs), and is required for growth. Copper deficiency can also lead to changes in coat colour: a red coat and the formation of 'spectacles' may be considered suggestive, and to compromised reproductive performance.

Due to its importance in so many areas and functions within the body it will come as no surprise that circulating levels of copper are kept under tight homeostatic

A red coat discoloration is frequently linked with 'copper deficiency', but are these calves suffering a primary deficiency of the mineral, or an interaction-induced secondary deficiency? Or is their copper status actually fine?

control. During periods where supply exceeds demand, copper is stored in the liver. It is then released to maintain circulating levels when supply is limited. This means that although blood copper levels can be estimated, a result within the expected range may fail to reflect adequately the true status of the animal. It may be that the release of stored copper from the liver is maintaining circulating levels of the mineral despite a deficient intake. By the time that blood levels fall below the expected range, stores of copper within the liver will have been depleted and productivity lost.

At the other extreme, storage of copper within the liver will maintain circulating levels within the expected range even if intakes are excessive. The capacity of the liver to store copper is, however, limited, and when this limit is reached, which can be precipitated by concurrent disease including liver fluke infection, fatty liver disease and toxic insult such as that caused by the pyrrolizidine alkaloids contained in ragwort and other toxic plants, the flood-gates open and much of the stored copper will be dumped into the bloodstream causing jaundice, haemoglobinuria and an often fatal haemolytic crisis. Obtaining biopsies of liver tissue (or samples from slaughtered animals) for laboratory examination may therefore be the best way of determining copper status, but this requires a surgical procedure, is invasive, and is not without risk (albeit small).

Add to this the interactions that occur between copper and iron (so why are so many mineral powders red, denoting an iron base?),

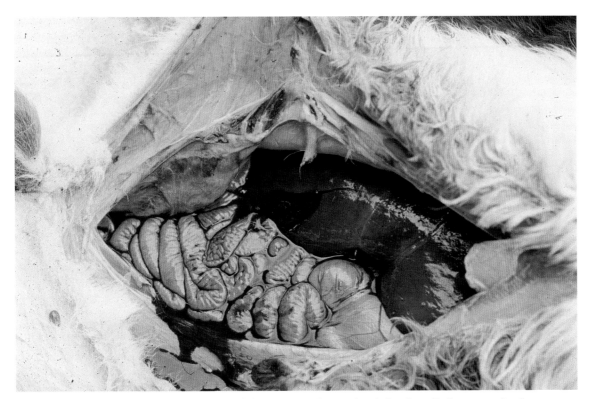

The typical bright yellow discoloration of the carcass of an animal that has died as a result of copper toxicity.

sulphur (which we spread liberally on some of our pastures), and particularly molybdenum, and the situation becomes even more complex.

Iron and sulphur can both combine with copper within the rumen forming insoluble complexes and reducing the amount of copper available for absorption. Furthermore, diets high in molybdenum also result in copper complexing to form tetrathiomolybdate, which also compromises the absorption of the mineral. All of this may contribute to an induced secondary copper deficiency. Molybdenum, however, is also thought to be toxic and has an adverse direct effect, particularly on reproductive function.

The debate between copper deficiency and molybdenum toxicity continues to rage. It may be better, therefore, to refer to copper-responsive disorders instead of either deficiency or toxicity. It is important, however, to differentiate between a primary copper deficiency and a copper-responsive disorder, because the means by which they need to be addressed may be different. Supplemental copper by almost any route should improve the situation in the case of a primary deficiency (but beware of inducing toxicity!), but in the case of a copper-responsive disorder it is the ratio of copper to molybdenum (and iron and sulphur) within the rumen that is important, so any necessary additional copper needs to be available orally.

Iodine

Iodine is an essential component of the thyroid hormones that control metabolic rate, foetal development and almost all bodily functions. Deficiency (although how this occurs when we use so many iodine-based disinfectants around many of our farms remains something of a mystery!) will, certainly in time, result in an enlargement of the thyroid gland in the neck, 'goitre', which may be particularly obvious in neonatal calves, can cause an increased prevalence of stillborn calves or late abortions, poor growth rates, low milk yield (which will further compromise calf growth rates in the suckler herd), changes to the skin and hair, and poor reproductive performance. An over-reliance on goitrogenic forage crops, often brassicas, may be significant in the aetiology of such problems.

Assessing iodine status, as with so many of the trace elements, is not necessarily easy or an exact science. When stillborn calves are available it is not difficult to dissect out the dumb-bell shaped thyroid gland

Dissection of a stillborn calf's ventral neck (head at the top of the picture) showing the dark, bi-lobed thyroid gland spanning the ventral aspect of the trachea just distal to the larynx. (The thymus gland is the larger, paler structure protruding from the cranial thoracic inlet lower in the image.)

(not to be confused with the thymus gland, which in neonates will protrude through the cranial thoracic inlet into the lower part of the ventral neck): it lies with the 'strap' across the ventral aspect of the trachea with one lobe to either side of it, just below the larynx, to be weighed, assessed for total iodine content, and examined histologically to assess follicular morphology. (Many complex formulae exist to determine whether the weight of the thyroid gland is excessive for any given calf, but as a rule of thumb, glands weighing less than 30g are likely to be normal.)

Blood samples can also be analysed for iodine content (plasma inorganic iodine – PII), although this process is expensive, is very susceptible to being artificially elevated (for example, if iodine-based disinfectants are used by the person taking the samples), and only provides an estimate of current iodine status when it is the level of thyroid hormones that is important. Alternatively, the circulating level of thyroid hormones, particularly T4, can be assessed, but these lag iodine intake by some weeks and reference ranges remain a matter for debate, having been changed dramatically in the relatively recent past.

Iodine supplementation is available in a variety of forms including boluses and free-access minerals or mineral licks, but splashing diluted iodine-based disinfectants on the backs or the flanks of the animals provides an empirical but often satisfactory means of supplementation, and adding potassium iodide crystals to the drinking water provides an even cheaper alternative.

Manganese

Manganese is required in very small amounts for the synthesis of mucopolysaccharides, an essential component of cartilage, and for the production of the steroid sex hormones, so deficiency has the potential to affect growth and fertility. In practice, however, such consequences are extremely rare.

In the suckler herd a deficiency of manganese during pregnancy in spring-calving cows housed over the winter and fed a grass silage-only diet has been linked with the birth of calves affected with congenital joint laxity and dwarfism (CJLD), often

A young calf showing signs typical of congenital joint laxity and dwarfism.

The same animal a year and a half later!

termed 'Acorn calves'. Such calves are typified by skull deformities (a superior brachygnathia – a shortening of the upper jaw), and shortened and abnormally bent limbs (chondrodystrophy) of varying severity, but an absolute demonstration of cause and effect has been difficult to achieve.

Selenium

Selenium, which can to some extent (but not completely) be compensated for by vitamin E when status is deficient, is required not only to promote optimal growth and fertility, but also in the formation of a wide range of seleno-proteins, with glutathione peroxidase (GSH-Px) being perhaps the most commonly considered and discussed; these are essential for a wide range of metabolic and immune functions. However, defining any particular condition as due to a selenium-deficient status can be challenging because of often multiple contributing factors, and so, rather than referring to deficiency diseases, reference is usually made to 'selenium-responsive disorders'.

Perhaps the best known consequence of selenium deficiency is white-muscle disease, a muscular dystrophy that predominantly affects cardiac and skeletal muscle, and which can, in the most severe cases, result in sudden death, particularly following exertion, as may be seen if the animals are gathered for handling. Impaired immune function as a consequence of a selenium-deficient status may present as an increased prevalence and severity of other infectious disease, pneumonia for example, and an adequate selenium status is required to ensure optimal early embryonic survival in breeding cattle.

Although tissue selenium content can be measured, circulating GSH-Px levels are usually used as an indication of selenium status; however, they lag intake by some time, so levels may remain within the expected range at times when intake is lacking, if historic intake has been satisfactory.

Zinc

Zinc is another mineral that is involved in multiple functions within the body. Deficiency (although how this can be an issue on many farms where galvanised metal is common, particularly as troughs to hold drinking water for the animals, presents a conundrum!) may contribute to poor immune function and presents as poor coat condition, parakeratosis (thickening of the skin), and the production of poor quality horn, resulting in an increased prevalence of lameness.

Assessment of zinc status requires the collection of blood to be tested into plastic tubes rather than the more usually used vacutainers with rubber stoppers. This is because the vulcanisation process employed during the manufacture of the rubber stoppers involves zinc, and this may influence test results.

Vitamin A

Vitamin A is not present in plant-based diets: it is synthesised within the intestinal wall and the liver from precursors, predominantly beta-carotene, and a synthetic form of the vitamin is added to most compounded feeds and supplements. Deficiency, which is usually only seen affecting intensively reared animals receiving inadequate supplementation, presents as blindness due to compression of the optic nerve due to an increase in intra-cranial cerebrospinal fluid pressure.

Vitamin B12

See 'cobalt' above.

Vitamin E

Vitamin E, in its various forms, is commonly present in both green forage and cereals, although it may – but not inevitably – break down during preservation. Deficiency is not considered common in suckler cattle, although intakes may be supplemented to help address marginal selenium status; *see* 'selenium' above.

WATER

The provision of an adequate supply of potable water to allow cattle to drink to maintain hydration, allow the rumen and other organs within the body, particularly the kidneys, to function optimally, and to ensure lactation and adequate nutrition for young calves in the suckler herd situation, as mentioned above, is vital to maintaining health and productivity. The importance of water as part of an adequate and properly balanced diet cannot be overstated.

MONITORING NUTRITIONAL STATUS

Metabolic Profiling

Metabolic profiling – collecting blood samples from representative animals within the herd and submitting them for laboratory testing – has become an established method of assessing dietary performance and metabolic status.

In most cases this testing can be carried out on serum obtained from clotted blood samples collected in red- or gold-topped blood tubes, or plasma obtained from blood samples collected in green-topped tubes containing heparin to prevent the blood from clotting. Certain analyses, however, require alternative anti-coagulants: glucose, for example, and phosphorus estimation are required to be carried out on samples collected in grey-topped tubes, and for accurate zinc status estimation tubes with plastic rather than rubber tops are required.

Beta hydroxy-butyrate and non-esterified fatty acid (NEFA) levels can inform on the energy status of the animals sampled, rising above the expected limit when dietary energy supply fails to meet demand and the shortfall is made up by mobilising body fat reserves. Estimating albumin and urea, or blood urea nitrogen, provides similar information about longer term and more recent protein status respectively. However, the interpretation of such profiles depends almost as much on the 'art' as the science. Circulating levels of beta hydroxy-butyrate and NEFAs will rise during periods when dietary energy intake fails to meet demand, as indicated above, due to the mobilisation of the body's fat reserves – but in cases of starvation, where the body's fat reserves have been totally depleted, they may return to the low levels that would normally be expected.

Low blood albumin levels might indicate a long-term protein shortage in the diet, but may be a consequence of liver disease compromising production, or of chronic kidney or gut disease resulting in a protein-losing nephthropathy or enteropathy; and a high circulating urea level might suggest a relative oversupply of RDP, but might also indicate renal compromise.

Mineral Status

Mineral status can be equally confusing. In some cases the mineral in question is measured directly – copper, for example – but this may not provide any information about bioavailability or activity, nor about confounding factors, including, in the case of copper, molybdenum, sulphur and iron status. In other cases, including selenium and iodine, status is usually estimated indirectly; for example glutathione peroxidase, a selenium-dependent enzyme, is usually measured to provide a proxy of selenium status and thyroxine, one of the thyroid hormones of which iodine is an essential component, may be used as a proxy for iodine status, both of which will provide an indication of historic rather than current supply.

In addition, the circulating levels of many, but not all, minerals are under tight homeostatic control, with minerals being stored within the body during times when supply is in excess to requirement, and being released from those stores when supply fails to meet demand, meaning that blood levels are maintained within tight boundaries until either body reserves have been completely deleted or are overwhelmed.

An alternative, particularly when assessing mineral status, is to assess the level of the mineral of interest in tissue samples; for example, liver samples are not uncommonly analysed to determine their copper and selenium status. To provide the most accurate information about herd status, however, this involves collecting biopsies from representative, healthy animals, which is expensive and invasive. It may also be possible to collect samples of the required tissue from animals slaughtered in the abattoir, although this will reflect status dependent on a finishing ration, which may differ considerably from the ration fed to breeding animals and the level of supplementation they receive.

Samples can also be collected from animals that die unexpectedly on the farm, either specifically to determine their status with respect to the mineral(s) of interest, or as part of a wider post-mortem examination (an estimation of kidney lead levels is often undertaken following a post-mortem examination to confirm or rule out lead poisoning as the cause of death).

Visual Appraisal

Having said all of this, your eyes also provide a very valuable (and cheap!) tool for the assessment of dietary adequacy: the cows and the herd should be critically appraised on a regular basis to note both appearance that may be suggestive of a dietary inadequacy, which can then be further investigated, and of body condition. Faecal scoring can also be used to provide an insight into the adequacy of the cow's digestive function.

Faecal Scoring

Faecal scoring simply involves assessing the consistency and content of the cow's faeces. As the cow defaecates the splash of the faeces hitting the floor should be reminiscent of a slow hand-clap, and a perfect cowpat, once formed, should stand proud of the ground with a small central depression in which it should be possible to 'stand a rose bud'! Any firmer or, particularly, looser, and something may be going wrong. Visible undigested

A firm faecal pat indicating good digestive function.

'Sloppy' faeces suggesting at least some degree of digestive upset.

grains or fibre also suggest a problem, either with diet preparation or with rumen function.

Rumen Fill

Rumen fill can be used as a guide to intake. This involves assessing the profile of the cow on the left-hand side of the abdomen, particularly in the left para-lumbar fossa just below the lumbar vertebral transverse processes. This area should appear full, but not over-full: if a large bulge is present it might indicate a degree of bloat, or that the diet being fed is of such low quality that digestion is impaired and consumption is increased to provide the required level of nutrition. If there is a significant dip inwards it suggests an inadequate dietary intake.

Body-Condition Scoring

Body-condition scoring was initially conceived to monitor the status of high-yielding Holstein dairy cows by giving a score, usually between 1 and 5 in the UK, with 1 representing emaciation and 5 representing obesity, based on the fat covering on either side of the tail-head and the prominence of the lumbar vertebral transverse processes. However, it can easily be adapted for use in beef suckler cattle. The exact condition score that is awarded to each

animal does not particularly matter: this is a subjective process and every scorer will differ. What does matter is that breeding animals in particular are neither emaciated nor obese, and consistency of estimation is important so that the process can provide accurate information about how the condition of the animals and the herd as a whole is changing.

An ideal situation in a spring-calving suckler herd might be for the cows to calve in a 'fit not fat' condition to minimise calving problems and maximise appetite immediately after calving to maintain energy status to promote uterine involution, a return to ovarian cyclicity, and milk production to feed the calf, coincident with the flush of spring grass growth. The cows should then be kept on a rising plane of nutrition throughout the serving period to maximise reproductive performance, and beyond; this is so that they go into the winter, when expensive conserved forage or, even worse, concentrates will need to be fed (there is no cheaper way of harvesting grass for cows to eat than letting the cows themselves graze it!), in the best possible condition to allow weight to be lost over the winter to achieve the desired body condition again at calving.

PROBLEMS ASSOCIATED WITH THE DIGESTIVE TRACT

Bloat

The production of gas, both carbon dioxide and methane, as described above, is an inevitable consequence of ruminant digestion thanks to the fermentation that takes place within the rumen. Under normal circumstances, this gas is voided to the atmosphere as the cow eructates (burps). If this does not occur, for whatever reason, the gas will build up within the rumen causing

BODY-CONDITION SCORING IN THE BEEF SUCKLER HERD

Body-condition scoring is not a new concept: stockmen have been critically observing the condition of their cattle for generations. The system of body-condition scoring currently in common use was developed and refined during the very late 1900s and early 2000s as a means of assessing the changes in body condition, and therefore the degree of negative energy balance (which has adverse consequences for production and future fertility) experienced by high-yielding Holstein dairy cows, particularly during the period after calving to guide nutritional management. It can be easily and usefully adapted for use in beef suckler cows with the same aim.

In the UK, body-condition scoring is usually carried out using a five-point scale between 1 and 5, with 1 representing emaciation, 5 representing obesity, and the other scores representing degrees of body condition between the two. Half scores are commonly given, with some assessors being even more detailed in their scoring and awarding quarter scores. Scores are, however, subjective, and it is not uncommon for different assessors to score the same animal at least half or even an entire condition score different. This is not necessarily important, assuming that the cattle are neither emaciated nor obese. What is important is consistency so that a picture of how body condition is changing, and therefore the plane of nutrition, can be built up.

The scores that are awarded are based on an estimation of fat cover over the hook and pin bones of the pelvis, on either side of the tail-head, over the lumbar transverse processes and over the ribs. Palpation as well as observation is important when carrying out body-condition scoring. Although precise definitions of the different scores can be given (as below), with experience, acceptably accurate scores, sufficient to guide the nutritional management of suckler cattle, can be rapidly attributed almost by eye without continual reference to these.

Body condition score 1: There is a deep cavity on either side of the tail-head. The lumbar spine, lumbar transverse processes and ribs are prominent and sharp.

What condition score best describes these cows? Should the thinner cow be given a BCS of 1 or 1.5? Should the fatter cow be given a BCS of 4, 4.5 or even 5? Does it really matter? These two animals are, of course, at the extremes of body condition, and yes, in cases such as these it *will* matter – but for most animals what really matters is how body condition is changing, and whether they are gaining or losing condition.

Body condition score 2: There is a shallow cavity on either side of the tail-head, and the pelvic bones remain prominent. The lumbar transverse processes are easily individually palpable, but with rounded ends. The ribs are individually identifiable but have some fat cover.

Body condition score 3: There is a small depression on either side of the tail-head but fat cover is present over the entire area. The lumbar transverse processes and ribs are only palpable with firm pressure.

Body condition score 4: There is obvious fat cover around the tail-head. The lumbar transverse processes and ribs are covered in fat and are not individually palpable.

Body condition score 5: The tail-head and other bony structures are 'lost' under a thick layer of fat.

the organ and the abdomen to distend, and raising the intra-abdominal pressure. If not relieved, this will eventually compromise venous return to the heart, and then, as the pressure rises still further, cardiac function resulting in circulatory failure and, in the most extreme cases, death.

Two types of bloat are recognised: frothy bloat and free-gas bloat. Frothy bloat is often associated with the inclusion of certain proteins in a rapidly fermentable ration; moving cows on to lush pastures with a high clover content is a particular risk. In frothy bloat the gas produced within the rumen is in the form of tiny bubbles creating the froth. Receptors at the oesophageal inlet detect this froth as a liquid and prevent the eructation reflex, resulting in the accumulation of more and more froth within the increasingly bloated rumen. Passing a stomach tube in such cases will often fail to resolve the problem. Instead, the frothy bloat must be converted into a free-gas bloat, which can then be relieved. This is done by drenching the affected animal(s) with an agent aimed at reducing the surface tension of the bubbles in the froth and allowing them to coalesce to form a discrete gas cap. Commercial 'anti-bloat' preparations containing poloxalene or other similar compounds are available, but a similar result can be obtained using vegetable oil or even detergent.

Free-gas bloat can arise for a number of reasons, including infection around the oesophageal inlet to the rumen inhibiting eructation, other systemic conditions (tetanus, for example), and oesophageal obstruction, with potatoes, turnips and apples providing common examples. It is usually a simple matter to relieve a free-gas bloat if a stomach tube can be passed, but if it proves impossible to pass the tube other action will be necessary. If a blockage in the oesophagus can be palpated before it disappears through the cranial thoracic inlet it may be possible to 'milk' the offending object back up towards and into

A calf with ruminal bloat that requires decompression either by the passage of a stomach tube, or more permanently by the placement of a 'Red Devil' (*see* image overleaf on the left) or the creation of a surgical rumen fistula (*see* image overleaf on the right).

Bloat relieved using a 'Red Devil'.

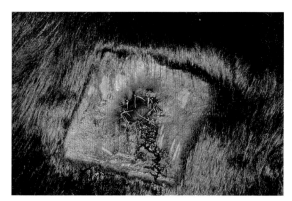

Bloat relieved by the surgical creation of a rumen fistula.

the mouth using your thumbs (although this is likely to cause cramp, so while giving your thumbs a rest it makes sense to pre-place a noose below the obstruction and tighten it every so often to prevent progress being lost).

More often, however, the blockage occurs within the chest as the oesophagus passes the heart. In these cases, if the cause cannot be gently pushed into the rumen using a probang or stomach tube (taking great care not to rupture the oesophagus) it may, assuming the obstruction to be of plant origin, be necessary to wait for it to soften

until it can be moved. Until this occurs it will be necessary to place an indwelling trocar or create a small surgical fistula through the left para-lumbar fossa and into the rumen to relieve the bloat and prevent it escalating again.

Acidosis

Acidosis is more a problem of dairy cows being fed energy-dense rations, often with additional concentrates being fed in the parlour to support high and persisting milk yields, than it is in suckler cows. However, where concentrate feed is provided to

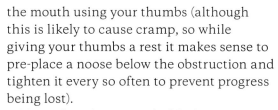

TRANSFAUNATION

Although many 'stomach powders' – often based on yeasts and ingredients including gentian, ginger and *nux vomica* – are available for the treatment of cattle with digestive upsets (and even feeding an armful of ivy might help), many will, at best, do little harm but may not have any great positive effect either. Transfaunation may, however, have a significant beneficial effect.

Transfaunation involves collecting several litres (a bucketful!) of rumen liquor from another healthy cow and then pumping it

into the rumen of the sick animal. A standard stomach pump can be used to collect the rumen liquor from a suitable restrained donor cow, although, obviously, it needs to be used in reverse, and the end of the stomach tube that is passed into the healthy cow's rumen needs to be weighted so that it falls through the mat of fibre and into the rumen fluid below it.

The rumen flora and fauna are very sensitive to changes in their environment, and they will start to deteriorate and die as soon as the rumen liquor has been collected and is exposed to the air, so transfer to the sick animal needs to take place without delay.

suckler cows in addition to the forage-based part of their ration, acidosis remains a possibility, and dramatic cases can present if cows break into concentrate feed stores and help themselves! In most cases, where the condition is mild, resolution is spontaneous, given time; the rumen buffering capacity of the saliva produced as the cows ruminate is more than sufficient to resolve most cases. In more serious cases providing fluids and an antacid by stomach tube (along with a NSAID, perhaps, to make the affected animal(s) feel better!) should hasten recovery.

A persisting low-grade acidosis often affects animals being intensively reared in 'barley-beef' units, despite buffers frequently being added to the diet. In these cases, assuming a slow transition on to the finishing ration to allow the rumen flora and fauna to acclimatise to what is being fed, acute problems are usually avoided – although such management systems have been associated with a higher than usual prevalence of abscessation, usually affecting the liver. This is assumed to be due to bacteraemic spread via the portal circulation, and although not usually fatal, will compromise performance to some extent and lead to rejections in the abattoir.

Rumen Impaction
Rumen impaction is a rare occurrence but can occur, as mentioned above, when a diet of such low quality is being fed that the rumen flora and fauna are unable to function adequately to digest the ingested fibre. As a result, the cow, although full, remains hungry, and continues to eat until her rumen can take no more. In most cases simply improving the quality of the diet will resolve the situation given time (although the loss of body condition during that time may be significant). In

more serious cases, as well as improving the quality of the diet, transfaunation (collecting rumen fluid from a healthy cow and then immediately pumping it into the rumen of the affected animal(s)) can help to hasten recovery.

Traumatic Reticulo-Peritonitis and Pericarditis ('Wire')
Cattle are indiscriminate feeders, ingesting mouthfuls of grass or conserved forage and swallowing it to be regurgitated and chewed later. This means that small contaminants in the grass or forage, short lengths of wire, for example, may be inadvertently swallowed along with the grass or forage. This is perhaps most commonly seen where degrading tyres that are shedding their radial wires are used to weigh down the plastic sheets used to exclude air from silage pits – although grazing over the site of a previous bonfire can also be high risk, as can the debris from newly erected fences if just discarded in the grass.

Once ingested, these short lengths of wire (classically an inch or two long) gravitate into the reticulum on the cranio-ventral aspect of the rumen where the usual contractions of the healthy gut may cause them to penetrate the reticular wall. At this time there will be an acute period of malaise which, particularly in the suckler situation, may not be appreciated. Following this acute illness as the wire penetrates the gut wall, a more chronic situation develops as the wire migrates through the cow's cranio-ventral abdomen (or further afield), causing infected tracts and abscessation. The liver is the most commonly affected organ, although it is even possible for the wire to penetrate the diaphragm, enter the ventral chest, and penetrate the pericardium around the heart and cause a pericarditis, ultimately

 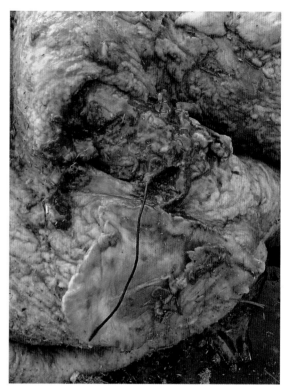

Pericarditis caused by a wire (seen in the image on the right) that has been ingested and has then migrated through the wall of the reticulum, the liver, the diaphragm and into the pericardium.

Degrading car tyres, used to weigh down silage sheets, are a common source of the wires involved in traumatic reticulo-peritonitis and pericarditis.

The author draining pus from the pericardium of a cow with traumatic reticulo-pericarditis. This will provide significant short-term relief, but the prognosis remains hopeless.

compromising cardiac function (traumatic reticulo-pericarditis).

Unfortunately, most of these cases are not appreciated until they become terminal, although occasionally, if the condition is recognised in its early stages, blousing the affected animal with a rumen magnet may drag the wire back into the lumen of the gut, and then a course of antibiotics may allow the infection to be brought under control.

Vagal Indigestion

Rumen contractions, which continuously mix the ruminal content and are necessary for successful digestion and eructation, are controlled by impulses travelling along the vagus nerve. If these impulses are compromised for any reason (perhaps an abscess that forms close to or around the nerve as a consequence of a case of traumatic reticulo-peritonitis), rumen contractions and digestion will also be compromised. This will

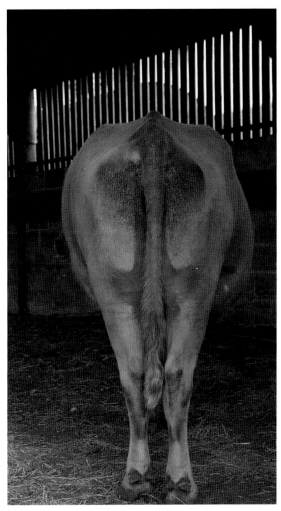

A cow displaying the typical '10-to-4' abdomen associated with vagal indigestion.

result in a loss of condition, a mildly bloated appearance, and a 'slippage' of the rumen within the abdomen from its usual vertical position on the left-hand side (when viewed from behind), resulting in bulges on the lower right-hand side and upper left-hand side of the abdomen – a so-called '10-to-4' appearance. Although such cases may appear to respond to antibiotic therapy, any improvement is usually limited and short-lived.

CHAPTER 3

BREEDING STRATEGY

Breeding strategy in the suckler herd, at its most basic, is aimed at delivering a calf per cow put to the bull each year. This, however, involves much more than the 'management by accident' approach of simply putting a bull with the cows and waiting: the timing of mating will be determined by the desired calving period and its duration, and the choice of the genetics used will be guided by the fate planned for the anticipated calves.

In the commercial suckler herd that is aiming to produce calves to be fattened and slaughtered for human consumption, cross-bred animals will be the usual choice for the breeding females, to maximise the benefit of hybrid vigour; if all else is equal the cows will usually conceive more easily, and the calves will grow better than pure-bred animals. Perhaps one of the best examples of this is the Stabiliser 'breed', a four-way cross first developed on the Leachman Ranch in California that combines Red Angus, Hereford, Simmental and Gelbveih: this cross produces a uniform, highly fertile suckler cow that maximises the advantages of hybrid vigour (although this will almost inevitably decline as the 'breed' becomes established and the gene-pool 'stabilises'). The pedigree herd obviously introduces different considerations, with the breed of animal often being dictated by tradition and personal choice. Breeding decisions within the breed, however, remain highly relevant and can have a significant impact on herd performance.

Cross-bred Blonde d'Aquitaine suckler cows on improved grazing.

Pedigree Sussex cattle.

NATURAL SERVICE

Despite the obvious and well-proven advantages conferred by hybrid vigour, it is usual practice in many parts of the world to run pure-bred bulls with suckler herds. The usual explanation for this is that when comparing calf performance, the genetic potential provided by different bulls of the same breed to their offspring can be analysed to select the most advantageous individuals to use to breed future generations; but this is not possible if cross-bred bulls are used. In such a situation it becomes impossible to determine how much calf performance is influenced by paternal genetic superiority, and how much is due to hybrid vigour. Hybrid bulls are, however, used to advantage by some farmers and breeders, perhaps most notably by Rick Pisaturo in Australia who, similarly to Lee Leachman in America, has developed various cross-bred animals – Mandalong Specials, Square Meaters and Tropicanas, for example – to sire superior calves that are suited to his local environment.

A Planned Approach to Breeding Strategy

Assuming a commercial suckler herd run in a traditional way with cross-bred cows and pure-bred bulls, there still needs to be a planned approach to breeding strategy. This might involve careful animal selection based on the planned management and feeding strategy for the cows, and the target market and fattening strategy for the calves.

There is no cheaper way of harvesting grass than for livestock to eat it, so there is no point in limiting access to grazing during the spring and summer so that grass can be set aside to cut for hay or silage (assuming that sufficient grass is available to harvest and conserve for winter feed) to be fed during the winter. It is also important to understand that suckler cows 'flush' just as sheep do, so it makes sense to calve suckler cows to coincide with the flush of grass growth in the spring: this will maximise the benefit obtained from forage, both in terms of promoting uterine involution and an early return to ovarian cyclicity after calving, and to optimise milk production to ensure the nutritional status and growth of the calf. Limiting the serving period, and therefore also the duration of the calving period, to twelve weeks or fewer, facilitates the management of the herd to realise this advantage.

Split calving patterns, with a proportion of the herd calving in the spring and the remainder calving in the autumn, may be justified based on labour availability or a more efficient use of bull power (assuming natural service) or, in the pedigree herd, by the need for bulls of various sizes and therefore ages to be available for sale; but such split patterns are difficult to justify in terms of managing grazing efficiently.

Pedigree Ruby Red Devon cattle grazing chalk downland in the south of England.

It is not, however, just when calving occurs and for how long the period extends that is important. When the cows calve within that calving period can also have a profound effect on herd performance and profitability. Calves born early within the calving period will be older and therefore heavier, and so more valuable at fixed-time weaning. Add to this the fact that infectious 'bugs' tend to build up as the calving period progresses, so early-born calves are less likely to become ill and suffer setbacks to growth than later-born calves, creating a 'double-whammy' situation.

In addition, cows that calve early during the calving period have longer to undergo uterine involution and return to ovarian cyclicity before they are put back with the bull than cows that calve later during the calving period if an annual calving pattern is to be maintained. Thus a cow that calves on the first day of a three-month calving period will have the entire three-month period before being put back with the bull; but a cow that calves on the last day of the three-month period will be put back with the bull on the following day: which cow will be most fertile when the bull is re-introduced?

It stands to reason that cows calving earlier during the calving period will therefore be more likely to conceive sooner after being returned to the bull than later calving cows, so cows calving early during the calving period will tend to calve early year after year, but cows that 'slip' and calve later during the calving period tend to slip more and more each year until they remain empty at the end of the serving

Housed cross-bred fattening cattle eating TMR (Total Mixed Ration).

period, thereby creating a dilemma: should the serving period be extended, and if so, for how long? Should the bull be removed at the end of the planned serving period but be reintroduced later in the year, resulting in a split calving pattern? Or should a 'perfectly good' cow be sold as barren, requiring an additional herd replacement? (This is often the real reason for a split or extended calving period, despite the often very plausible explanations that are given!)

As a rule of thumb, you should target to calve at least 60 per cent (and hopefully more!) of the herd within the first 'cycle' (three weeks) of the calving period, or half of the herd within the first eighteen days to optimise production efficiency.

Size of cow and effectively, therefore, breed also needs consideration based on grazing management and the quality of the grazing available, and the breed of the sire needs consideration based on the planned fate of the calves produced. Smaller cows tend to have a lower maintenance requirement than larger cows, and traditional native breeds have a larger ratio of gut volume to mature body mass than many larger continental breeds; for this reason they will often perform better on poorer quality pastures and if kept outside for longer, particularly during periods of inclement weather, or even year-round. Similarly, calves sired by native-breed bulls will usually be better suited to fattening outside off grass than those sired by continental bulls, which may be better suited to fattening indoors on a diet of conserved forage and concentrates (or almost exclusively concentrates).

Estimated Breeding Values

Estimated breeding values (EBVs) give a guide to the potential of the offspring of any bull being considered, whether a natural service sire or one from the AI stud – but beware! EBVs are just that: a guide – although reliability will increase as information is gathered about the performance of a bull's offspring to replace information deduced from his pedigree; information known about related animals should also be considered.

Calving ease is always important, of course, but other EBVs should be considered depending on the intended fate of the offspring. When considering bulls to sire animals to be fattened, growth rate and fat cover EBVs, amongst others, will be important, but when selecting a bull to breed replacement breeding heifers, gestation length and milking ability will assume a greater importance. Using bulls that pass on a shorter gestation length than breed average will, over many generations, result in dams that carry their calves *in utero* for a shorter period of time, leaving a longer 'open' period after calving for uterine involution to occur and the animal to return to cyclicity before she needs to conceive again to maintain an annual calving pattern. This maximises the probability of conception within the desired period, and using bulls with a higher milking ability than breed average will ensure that their daughters, if retained as breeding animals, will produce sufficient milk to feed and grow their calves.

When the aim is to produce heifers to be retained to join the breeding herd, using bulls with a large scrotal circumference is also advantageous, as this is correlated not only with their own fertility but also with the fertility of their offspring. It is important to understand, however, that as well as only providing a guide to the potential of their offspring, EBVs are only of use to compare the potential of the offspring of different bulls within a single breed.

Charolais cows and calves at summer grazing.

ARTIFICIAL INSEMINATION (AI)

Although reproduction, certainly in the commercial suckler herd, almost always relies on natural service, the pros and cons of artificial insemination (AI), and the advantage that the superior genetics available through the AI stud brings, should also be considered. The cost of acquiring and maintaining the best natural service sires can be considerable, accidents may happen, their fertility cannot be guaranteed, and the health and safety aspects of keeping such a large animal (he does not need to wish to hurt you: simply playing with you can lead to serious injury, or worse) all need to be considered.

Using AI removes many of these potential dilemmas while allowing easy access to a choice of the best genetics available worldwide at very reasonable cost, such that different sires, several if wished, can be used across the herd targeted at producing offspring that in turn are targeted for different fates from different cows. It does, however, introduce the challenge of ensuring that cows are served at the optimum time to maximise conception and pregnancy rates. This can, of course, be done by carful observation of the herd, assisted by oestrus detection aids, to identify animals showing the typical mounting behaviour that indicates a 'bulling' animal, then serving the cows individually at the appropriate time.

Although good results can be achieved when operating such a system, it does require excellent stockmanship and a significant amount of time: the herd,

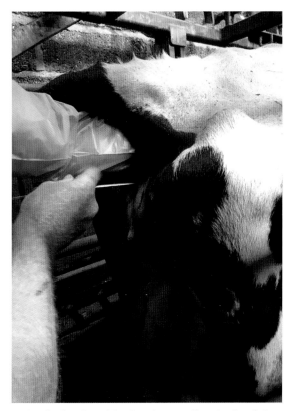

AI in the beef suckler herd, as well as in the dairy herd, allows access to a wider range of often superior genetics than is commonly available using natural service.

depending on its size, should be viewed for a minimum of 20 or 30 minutes several times each day to maximise the chance of identifying as many cows as possible as they come into oestrus. An alternative approach is to synchronise the oestrus cycles of groups of animals so that the timing of oestrus can be predicted and they can all be served together. A variety of 'programmes' using exogenous hormone treatments is available to achieve this, but there are inevitably costs, and again, time implications and the success of such programmes in achieving pregnancies can vary. Nevertheless, assuming good nutritional management and the control, or absence, of infectious disease, it should be possible to repeatedly get two-thirds or thereabouts of groups of synchronised animals pregnant to fixed-time AI.

Irrespective of whether service is to observed oestrus or following oestrus synchronisation, sympathetic handling of the cows through an efficient handling system will be necessary to reduce stress (to both cows and stockmen!) and maximise results.

SUGGESTED KEY PERFORMANCE TARGETS

The following are suggested key performance indicators (KPI) for seasonally calving UK suckler herds using natural service.

Bull:cow ratio	Depending on age
Duration of serving period	10 weeks (certainly <12 weeks)
Pregnancy rate	>95 per cent
Cows aborting after being confirmed pregnant	<2 per cent
Cows calving during the first three weeks of the calving period	65 per cent
Cows calving during the second three weeks of the calving period	25 per cent
Cows calving during the remainder of the calving period	5 per cent
Calves stillborn or dying before reaching 24 hours of age	<2 per cent
Calves weaned (of those alive at 24 hours of age)	>98 per cent
Calves sold (of those weaned)	>98 per cent

REPLACEMENT POLICY

Replacement policy can also have a significant impact on suckler herd performance and profitability. Cows will, of course, have to leave the herd as they become older and inefficient, but these 'cull' cows should be regarded as a resource to provide an additional source of income: they should be managed so that if destined for slaughter for human consumption they leave the herd in the best possible condition.

These cull cows will need to be replaced with younger animals if the size of the herd is to be maintained – but should these replacement breeding animals be purchased or home-bred? Home-bred animals have the advantage of known potential and health status, but if home-bred heifers are retained, the number of calves produced from the herd for sale or fattening each year will be reduced. Producers should also beware of declining milk production from heifers bred from many generations of beef genetics, which may eventually have an adverse effect on calf growth rates, irrespective of potential.

Purchased animals carry exactly the opposite dilemmas, and ensuring health status and managing effective biosecurity can be major challenges. However, milk production will not be an issue if the animals selected are cross-bred females from the dairy herd.

If home-bred animals are to be retained as replacement breeding animals a decision then needs to be made about how many and which. Most commercial suckler herds will require about fifteen replacement breeding females each year for every one hundred cows in the herd. Assuming that all these cows calve each year, that all the calves are born alive and none of them die – possible, but perhaps unlikely! – there will be a pool of fifty heifer calves each year from which to select your replacement breeding animals. Some of these, the later born and therefore smaller calves perhaps (another reason to maintain a tight calving pattern!), will obviously not be big enough to be served to deliver their first calf at two years of age, but there are likely to be thirty-five or forty animals from which the fifteen required must be chosen. This choice may then be based on maternal performance, conformation and personal preference.

However, if fifteen animals are selected, assuming a conception rate of 60 per cent, it will take four bulling cycles – twelve weeks, which is often the entire duration of the planned serving period – to get fourteen of them pregnant. Why not put all thirty-five or forty to the bull at the desired time? In this situation, with the same 60 per cent conception rate, the required fifteen heifers will be pregnant in less than the first three weeks bulling cycle and the bull can be removed, resulting in natural selection for fecundity and in early calving heifers joining the herd, maximising their potential life-time production. (The heifers that do not conceive can then still be fattened or sold for fattening for human consumption, and if more heifers than are needed are found to be pregnant, these, too, can be sold as replacement suckler cows to be introduced into other herds.)

CHAPTER 4

MANAGING FEMALE FERTILITY

Female cyclicity, including ovarian function and ovulation, oestrus behaviour and maintenance of the uterine environment to ensure that it is adequate to receive and support the fertilised egg, is driven by complex interactions and feedback loops involving multiple hormones acting within the hypothalamic-pituitary-ovarian axis.

THE OESTRUS CYCLE

Gonadotrophin-releasing hormone (GnRH) is released from the hypothalamus in a pulsatile manner, stimulating the release of follicle-stimulating hormone (FSH) and lutenising hormone (LH) from the anterior pituitary gland under the control of the ovarian steroid hormones oestradiol, which is produced by the maturing follicle, and progesterone, which is produced by the corpus luteum. FSH acts to recruit new follicles to develop. Development is maintained by the influence of pulses of increasing magnitude of both FSH and LH due to positive feedback on the hypothalamic-pituitary axis of oestradiol from the developing follicle, until a surge in LH stimulates ovulation and oestrus behaviour.

Following ovulation, the ruptured follicular tissue remaining in the ovary becomes the corpus luteum, producing progesterone: this exerts a negative feedback on the hypothalamic-pituitary axis, suppressing

further follicular development and preparing the uterine environment to receive the fertilised egg and support a pregnancy. If the egg is not fertilised or the pregnancy is not successful, prostaglandin F2a produced within the uterus acts to cause luteolysis, and the cycle, which in the cow takes approximately twenty-one days to complete, begins again.

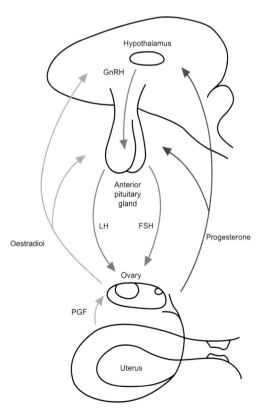

The hormonal feedback loops governing the hypothalamic/pituitary/ovarian axis.

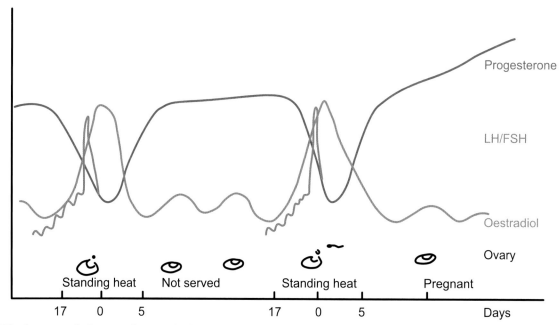

Progesterone

LH/FSH

Oestradiol

Ovary

| Standing heat | Not served | | Standing heat | Pregnant |

| 17 | 0 | 5 | | 17 | 0 | 5 | Days |

The hormonal changes during the bovine oestrus cycle and the early part of pregnancy.

The cow, having reached puberty, will cycle continuously year round, unless or until pregnancy intervenes, assuming an adequate body condition and plane of nutrition, and unless prevented by disease or pathology.

OESTRUS SYNCHRONISATION PROGRAMMES

All synchronisation programmes used to manipulate bovine reproduction rely on removing progesterone, either endogenous progesterone from the corpus luteum, or exogenous progesterone from a progesterone-releasing device usually inserted into the vagina, to bring the cow into oestrus and cause the dominant follicle to ovulate at a predictable time. The removal of endogenous progesterone is achieved by the administration of prostaglandin F2a (PGF), which requires the cow to be cycling normally and

the presence of a corpus luteum to be successful.

The advantage of a synchronisation programme relying on the removal of exogenous progesterone, which is achieved simply by removing the progesterone-releasing device being used, is that cows do not necessarily need to be cycling normally for success to be achieved. Anoestrus cows, perhaps because they have not yet returned to cyclicity following a recent calving or perhaps because of a slightly poor body condition, can be induced into oestrus and to ovulate using programmes relying on the provision of exogenous progesterone, although conception may not be maximal in such situations.

Once the basic programme has been decided it is then possible to manipulate it in a number of ways with additional treatments, particularly involving the administration of GnRH (gonadotrophin-releasing hormone) at various times to synchronise follicular waves and recruit follicles to mature to produce a dominant follicle, and to ensure the ovulation of that follicle at the appropriate time to maximise the chance of fixed-time service resulting in a pregnancy.

It is important to note that for the success of any synchronisation and fixed-time service programme to be maximised the cows or heifers involved must have good uterine health, be in 'fit not fat' body condition, and on a rising plane of nutrition with all their mineral requirements met and all relevant infectious diseases absent or adequately controlled.

Double-Dose PGF (Prostaglandin F2a)

A double-dose PGF synchronisation protocol works well in animals that are cycling normally. It does not matter that the animals selected will be at an unknown and largely random stage of their cycle when the first dose of PGF is given; those in which a corpus luteum is present will undergo luteolysis as a result of the first treatment, and although there will be no effect in those in which a corpus luteum is not present, all the animals in the group will now be at about the same stage of the ovarian cycle and so will progress to develop a corpus luteum at about the same time. Eleven days later all the animals in the group would be expected to have a functioning corpus luteum producing progesterone, and so a second dose of PGF at this time will synchronise luteolysis, oestrus behaviour and ovulation amongst the group of animals.

As in any biological system some variation in response to treatment can be predicted, so to maximise success, synchronised animals should be served twice, three and four days after the second dose of PGF. However, satisfactory results can often be obtained following a single service about three and a half days after the second dose of PGF (the exact timing will depend on the animals being synchronised; heifers usually respond

Double dose PGF 2a oestrus synchronisation and fixed-time AI

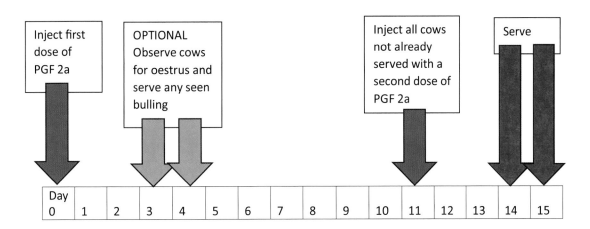

Oestrus synchronisation for fixed-time AI using a double-dose prostaglandin programme.

slightly faster than cows), with consequent savings in the number of times the animals need to be handled, and semen costs if artificial insemination is being used.

A major advantage of such a synchronisation programme, assuming adequate handling facilities and the presence of experienced stockmen who can detect oestrus cows, is that any animals that are seen 'bulling' following their first dose of PGF can be served to this observed oestrus (only a single serve is required in this case), and then removed from the process. This will provide further savings in terms of cost and time, although greater management input will be required.

Synchronisation using Exogenous Progesterone

In programmes using exogenous progesterone, the latter is usually supplied by inserting a progesterone-releasing intra-vaginal device, a PRID or a CIDR. (Progesterone-impregnated silastic rubber ear implants, Crestar, which were positioned subcutaneously at the back of the ear to achieve the same objective, are no longer available in the UK.) The implant is then removed after remaining in place for an appropriate period of time, usually between seven and twelve days in beef cows. (Shorter periods are often used in dairy cows, particularly Holstein cows, reflecting differences in follicular dynamics between beef and dairy cattle, and in beef suckler heifers reflecting the difference in metabolic demand experienced by lactating cows and heifers being served for the first time.)

However, the precipitous drop in circulating progesterone levels required to achieve oestrus synchronisation will only be achieved if no source of endogenous progesterone, luteal tissue, is present, so to maximise synchrony, it is also necessary to give a dose of PGF at, or ideally twenty-four hours before, implant removal. Again, to maximise success, synchronised animals should be

Progesterone oestrus synchronisation and fixed-time AI

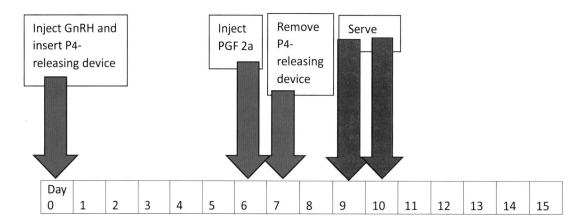

Oestrus synchronisation for fixed-time AI using an example progesterone-based programme.

served twice, but in this situation two and three days following implant removal. However, a single service between fifty-six and sixty hours after implant removal should, again, provide satisfactory results.

All other synchronisation programmes are essentially variations of one of these two programmes, trying to provide refinement, greater synchrony, and greater success in achieving pregnancies following fixed-time service. The benefits that may be accrued from these variations need to be weighed carefully against the increased number and cost of the hormonal treatments involved, and particularly the time and stress to the cows that is involved in the increased number of handlings that they entail.

Even without embarking on an enhanced synchronisation programme, assuming that cow condition, nutrition and infectious disease control are all optimal, pregnancy rates of between 55 and 65 per cent following oestrus synchronisation and fixed-time service can be expected (although this can vary widely). The herd manager will therefore need to consider how the synchronised cows that fail to conceive following their fixed-time service will be managed. Many farmers will follow oestrus synchronisation with fixed-time artificial insemination (AI), and will then introduce one or more bulls so that any cows that do not conceive to the AI can be served again naturally as they return to oestrus; it may be confidently predicted that at least half of the synchronised cows will have conceived to AI, and bull requirement will therefore be calculated based on this.

Do not forget, however, that although the synchronicity of the cows' cyclicity will drift the longer the time period that has elapsed following treatment, those that have not conceived will still tend to return to oestrus within a relatively short period of time. This means that more bulls may be required than was predicted based on the number of cows that will still require serving, but which are cycling randomly.

FERTILISATION AND THE ESTABLISHMENT OF PREGNANCY

Ovulation of the dominant follicle usually occurs towards the end of standing oestrus, releasing the mature oocyte into the abdomen from where it migrates down the fallopian tube, assuming tubal patency (this can be assessed, if required, by 'dye' testing) towards the uterus.

TUBAL PATENCY ('DYE') TESTING

Where a cow is cycling regularly, with a normal inter-oestrus interval, and yet repeatedly fails to conceive despite being served by a known fertile bull or competent AI technician, damage and blockage of the fallopian tubes, which will prevent the sperm and eggs meeting and fertilisation occurring, should be included on the list of possible differentials. After other possibilities for the failure of the animal to conceive have been ruled out, this possibility can be investigated by carrying out a tubal patency or dye test.

This involves catheterising each uterine horn in turn using a foley catheter, before infusing a harmless dye, Phenolsulphonphthalein (PSP), which is excreted by the kidneys. If the fallopian tubes are patent, dye will flow up them into the abdomen, be absorbed into the bloodstream, be excreted by the kidneys and become visible in the urine. An absence of dye in the urine after an appropriate period of time confirms blockage of the fallopian tubes.

Fertilisation occurs within the fallopian tube towards its junction with the uterus. Sperm cells are deposited in the anterior vagina following natural service, or in the body of the uterus just cranial to the cervix if the cow is artificially inseminated, and then move *en masse* through the cervix (following natural service), through the body of the uterus, along the uterine horns, and then enter the fallopian tubes as a result of both passive transport and active propulsion. During this journey they undergo a process called capacitation, and structural changes to the membrane covering the acrosome to allow fusion with, and penetration of, the membrane surrounding the ovum.

This process takes approximately six hours, but does vary: some bulls produce sperm cells that capacitate early, and some produce sperm cells that capacitate late. (This is taken advantage of in 'fertility-plus' AI straws, where sperm cells from bulls that capacitate at different times are combined to try to ensure that at least some fully fertile sperm cells meet the ovum at the optimum time to ensure fertilisation, irrespective of the exact time during standing oestrus that ovulation occurs, and how long it then takes the ovum to travel down the fallopian tube.)

Fertilisation is achieved when usually a single sperm cell joins with the ovum to form a new diploid cell when the chromosomes from both join.

Following fertilisation, the new diploid cell starts to divide rapidly, forming first a solid, multi-celled morula. Cellular differentiation then begins – though one might ask how, when every cell in this very early embryo is genetically identical. The reason is epigenetics, although the science of this is still poorly understood; although all the cells in the early embryo are genetically identical they are not all exposed to the same environmental influences because of their position within the cellular mass.

This cellular differentiation results in the formation of a blastocyst with a peripheral trophoblast, which goes on to form the placenta and foetal membranes; an inner cell mass within which cells differentiate to

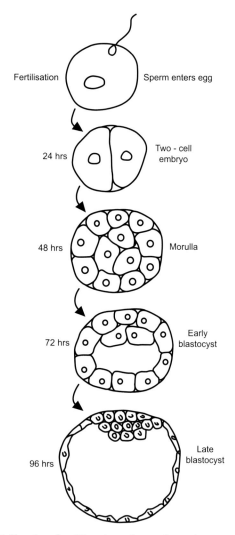

Following fertilisation, the early embryo undergoes rapid cell division and growth, becoming first a solid ball of cells, a morula, which then, in the early blastocyst stage, develops a lacuna, and then, in the late blastocyst stage, differentiates into trophoblast, which goes on to form the placenta and foetal membranes, and the inner cell mass that forms the embryo.

form ectoderm, from which the skin, hair, mammary tissue and nervous tissue are derived; mesoderm from which the bones, skeletal muscle and cardiac muscle are derived; endoderm from which the other internal organs are derived; and a fluid-filled blastocoele. By about a week after fertilisation the fertilised ovum enters the uterus, 'hatches' from its membrane, the zona pellucida, and begins a period of rapid growth.

If fertilisation has not been successful, or if the cow has not been served, luteolysis begins just over two weeks after ovulation, progesterone levels start to fall, and the cow will return to oestrus. To establish a successful pregnancy this must be prevented. This relies on the production of an interferon-like protein, bovine trophoblast protein 1 (bTP1), by the early embryo from about ten days after fertilisation, depending on embryonic development: this signals the presence of the embryo and ensures persistence of the corpus luteum to continue to ensure the progesterone dominance required for the establishment and maintenance of pregnancy. (In early pregnancy almost all the circulating progesterone will be provided by the corpus luteum, but later, progesterone from other sources, particularly the placenta, can be sufficient to maintain the gravid state.)

A failure of the embryo to produce bTP1 either early enough or in sufficient quantity can result in early embryonic loss. Although many other causes for early embryonic loss also exist, some early pregnancies may be saved by the administration of a dose of exogenous GnRH (gonadotrophin-releasing hormone) on day ten or eleven after service, buying the early embryo a little more time to adequately signal its presence.

Implantation, assuming a satisfactory uterine environment, takes place about a month after fertilisation, with, in the bovine, the development of the cotyledonary placenta to provide the developing embryo and foetus with oxygen and nutrition.

PREGNANCY DIAGNOSIS

Pregnancy diagnosis (PD), or PDing, perhaps alongside body-condition scoring and bull breeding soundness examinations, is one of the most important tasks necessary to guide management decisions in any cattle-breeding enterprise, and certainly in beef suckler herds. It can be achieved by a variety of methods, including the measurement of hormone level fluctuations, or the presence or absence of specific proteins at a set time after mating. In the beef suckler herd, however, the most convenient methods are *per rectum* palpation or ultra-sound examination of the cows' reproductive tract.

Pregnancy diagnosis provides information that is vital to making informed decisions about suckler herd and individual cow management.

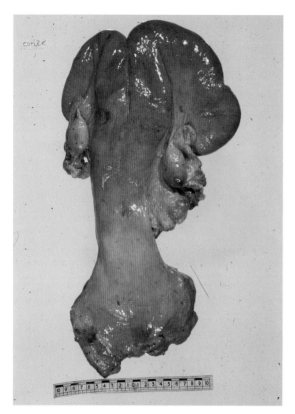

A gravid uterus containing an early pregnancy.

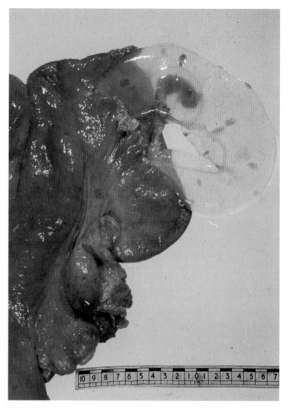

The early pregnancy.

Rectal Palpation

Rectal palpation can be used to detect pregnancies with acceptable accuracy from as early as four to six weeks after mating depending on the skill of the operator, the parity of the cow and her body condition: the older and fatter the cow, the more difficult it becomes to detect early pregnancies. The task involves careful palpation of both uterine horns, which aims to detect either cotyledons or a calf's head in more advanced pregnancies, or in the case of earlier pregnancies a discrepancy in uterine horn size and tone with a typical vibrant fluid feel of the larger, pregnant horn, and 'membrane-slip' as the uterine wall is gently pinched and the foetal membranes 'ping' between the thumb and fingers. (Great care needs to be taken when PDing cows expected to be during the early stages of pregnancy to avoid

damaging the embryo resulting in its death, and to avoid damaging the uterine wall and causing the release of prostaglandins, which may cause luteolysis and the subsequent loss of the pregnancy.)

Ultra-Sound Scanning

Ultra-sound scanning has the advantage of allowing the operator to visualise the developing embryo (or later, the foetus and placenta) earlier than it may be possible to detect the pregnancy by palpation, and with less risk to the pregnancy. It is also possible to assess the viability of the embryo from as little as a month or even less after mating using an ultra-sound scanner by visualising a beating rudimentary heart. However, the more the boundaries are pushed, and the earlier that pregnancy diagnosis is carried out, the greater the risk that a cow that is found to

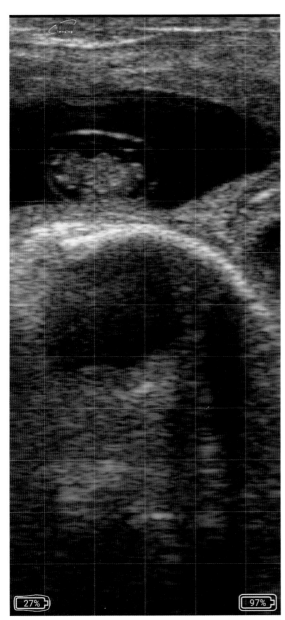

An ultrasound scan confirming pregnancy
34 days after serving.

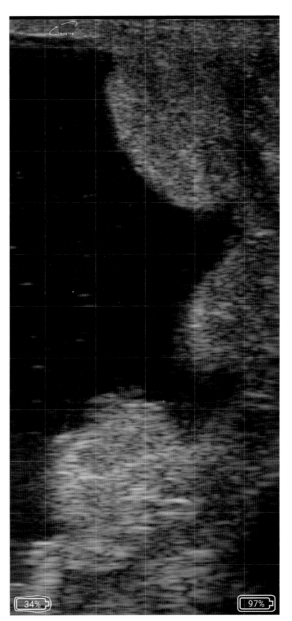

An ultrasound scan of a later pregnancy showing
placental development and the presence of
cotyledons.

be 'in calf' will lose the pregnancy for one of a
multitude of reasons following the PD; if this
is not noticed (and there may be little to notice
other than perhaps a return to oestrus if an
early pregnancy is lost) it will result in the cow
failing to calve at the expected time, or at all.

PROBLEMS

Pregnancy Failure, Early Embryonic Loss and Abortion

A multitude of causes exists for the loss of
an early pregnancy: a genetic composition

incompatible with life; a failure of the early embryo to produce sufficient bTP1 early enough to signal its presence to the dam and prevent the usual luteolysis that occurs in non-pregnant cows; an inadequate uterine environment, perhaps as a consequence of infection or pathology, that fails to support the developing pregnancy; or iatrogenic caused by the administration of exogenous prostaglandin following the mistaken diagnosis of a pregnant cow as 'empty'.

In many suckler herds, particularly those relying on natural service, any early embryonic loss and its extent may not be obvious. While affected cows will often have an extended inter-oestrus interval, they will often conceive again within the same bulling period and be found to be pregnant after the bulls have been removed from the herd. However, losses can mount up in terms of an extended calving interval and an increased number of cows remaining empty at the end of the serving period.

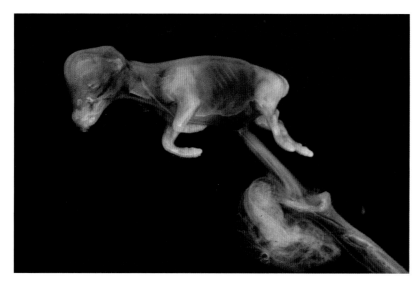

An aborted early embryo. The affected cow in this case may well return to oestrus, albeit after an unusually long period, when it will be served and conceive again.

A later abortion, which, in the suckler herd, will often result in the cow remaining unproductive and sold 'barren'.

Abortion refers to later foetal loss and the delivery of a dead or non-viable calf from about six weeks and before 271 days after service. Within Great Britain this remains a notifiable occurrence, and in a suckler herd is likely to trigger statutory testing for brucellosis as part of the national surveillance programme for this infectious agent. In practice, however, many abortions, and certainly early abortions, within suckler herds are not noticed until a cow that has been PDed 'in calf' fails to calve at the expected time.

In addition to brucellosis there are many other causes of cows aborting, including a plethora of viral, bacterial and protozoal infectious agents, fungi, toxins, compromise to nutritional status (including mineral status), and 'stress'. Investigating the cause, above and beyond any statutory testing that may be required, can provide useful information about herd health status, and which significant infectious agents may – and just as importantly, may not – be present to guide herd management.

ABORTION INVESTIGATION

Within the UK it is required by law to report bovine abortions to the competent authority (APHA – the Animal and Plant Health Agency), who will decide whether sampling and testing to rule out brucellosis is required. This is vital to ensure adequate surveillance in order that confidence in the national disease-free status can be maintained. If sampling is required, the aborting cow should be isolated from the herd and maternal blood and milk samples and a vaginal swab collected and submitted to an appropriate diagnostic laboratory. If available, samples of foetal fluid and foetal stomach content may also be useful.

If further investigation is required, the best chance of obtaining a positive diagnosis of the cause of the loss (although this is still not high) is to submit the foetus and placenta to an appropriate diagnostic laboratory as soon as is possible. Maternal blood samples may also be useful, though often more to rule out possible causes for the abortion, than to define the cause with certainty; just because a cow has been exposed to a particular pathogen and has seroconverted to it does not necessarily confirm it as the cause of the abortion (although it may give a useful indication of the health status of the herd, which may then prompt management intervention).

If it proves impossible to deliver the entire foetus and placenta to a diagnostic laboratory it will still be possible for a range of tests for the more common causes of bovine abortion to be carried out if selected tissues can be collected and submitted. The tissues recommended for submission include maternal blood, fresh and formalin-fixed placenta (including one or more cotyledons), foetal fluid, foetal stomach content, and a range of fresh and formalin-fixed foetal tissues including particularly, if possible, brain stem, thyroid gland, cardiac septum, lung, thymus gland, liver, kidney and spleen, along with any tissue with an abnormal appearance. Some of the more common causes of bovine abortion that might be worth investigating include iodine deficiency, BVD, BoHV1, Schmallenberg and blue tongue viruses, salmonella, leptospirosis, listeria, campylobacter, neospora, coxiella and fungal infection.

It is important that all aborted material to be delivered to the laboratory is packaged and labelled according to the relevant regulations, and that the remains of the aborted foetus and placenta are disposed of also in compliance with regulation.

The immediate losses to suckler herd productivity as a consequence of cows aborting is obvious: the absence of a live calf means one fewer weaned animal to be sold or retained as a future breeding animal. However, more subtle losses can also occur. Cows that abort a dead foetus may be more likely to 'hold their cleansing' than a cow that delivers a single, live calf at term, and as a result suffer subsequent endometritis or metritis. This will, in turn, likely delay uterine involution and the return to ovarian cyclicity, resulting in a prolonged calving-to-conception interval and calving index – if, indeed, they do conceive within the subsequent desired service period. Cows that abort, therefore, even if not deemed as 'barreners' by choice, are at greater risk of having to leave the herd because they fail to conceive within the desired period after aborting.

Congenital Conditions Affecting Female Reproductive Performance

A range of congenital or developmental abnormalities exists that will have an adverse impact on female reproductive potential, either directly or indirectly, or will prevent it altogether. These include, for example, segmental aplasia or a lack of patency of part or all of the reproductive tract as is seen in 'white heifer disease', a recessive genetic condition linked to white coat coloration and seen particularly in Shorthorn and Belgian/British Blue cattle. Perhaps the most common congenital abnormality affecting female reproduction in cattle is Freemartinism.

Freemartinism

In Freemartinism the developing foetus will, by default, develop as a female unless male hormones are produced to cause it to develop as a male. Where a twin pregnancy is present in the limited space in the uterus, it is possible for the two placentas to grow together and coalesce, resulting in blood passing between the two calves and for them to be born chimaeric. Where the twin foetuses are of the same sex, this is irrelevant in terms of development as either a female or a male, but where there are mixed sex twins it allows the passage of male hormones from the male foetus to the female foetus, causing abnormal development. This provides a diagnostic opportunity, because if a blood sample collected from the female calf contains both 'XX' (female) and 'XY' (male) cells, it confirms the calf as a chimaera and

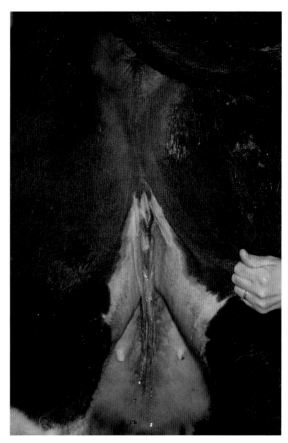

The abnormal external genitalia of a 'Freemartin' heifer (a female calf born twinned with a bull).

therefore the passage of blood containing male hormones between the calves. It also confirms that the female calf will be a Freemartin.

The extent of the abnormalities seen in Freemartin calves varies widely: some will be missing almost all of their reproductive tract, or one or both uterine horns may be imperforate, with a vestigial vagina; others, however, may be almost entirely normal females, and a small proportion of the female calves twinned with a bull calf will be able to breed. This depends on when during foetal development the two placentas join and for how long during gestation the female foetus is exposed to hormones from the male calf, as well as the magnitude of the exposure. How early the placentas join depends, to some extent, on where the twin embryos implant within the uterus: the closer they are to each other, the earlier their placentas are likely to join – so a female foetus that implants in the same uterine horn as a male foetus is at greater risk of being more severely affected than a female foetus that implants in the opposite uterine horn to a male foetus. The earlier the placentas join, and the more extensive the degree of coalescence, the more significant the abnormality is likely to be.

Acquired Conditions Affecting Female Reproductive Performance

Acquired conditions that may have an effect on future female reproductive performance can, as usual, broadly be divided into traumatic and infectious in origin.

Damage Caused by a Traumatic Calving

Calving is a traumatic process involving the relaxation and stretching of multiple tissues in and around the pelvis to allow the passage of the calf from the uterus to achieve independent life. Where problems occur, perhaps through a lack of patience and over-zealous assistance, perhaps because of maternal conformation or condition, or perhaps because of a foetal malpresentation or a foetus that is relatively or absolutely over-size, damage to the female reproductive tract is likely, if not inevitable. It is the site of this damage and its extent that dictates what action, if any, should be taken, and the consequence it may have on the future reproductive potential of the dam (or even her life). Small tears in the vaginal wall or vulval lips are frequently of little concern; the capacity of the female reproductive tract to heal is phenomenal (although trauma does present an opportunity for infection to become established, and this may need addressing).

More significant damage, even if it heals adequately, may lead to an incompetence at the vulval orifice and an animal that 'windsucks', resulting in continuous contamination of the vagina, which may affect future conception. (Surgery is possible to address this and is commonly carried out in mares with similar acquired deficiencies, where it is known as 'Caslick's procedure'.) In the most severe cases there may be complete disruption of the perineum, creating a union between the vagina and the rectum, allowing faeces to pass between the two. In such cases, although healing is usual, the persistent presence of faeces in the vagina will obviously not have a beneficial effect on the establishment of pregnancy (although somehow it does not preclude it in every case), and again surgery is indicated to reconstruct the hind end of the cow.

Where this is attempted it is often preferable to wait until initial healing has occurred and the surrounding stretched and bruised tissues have returned to a more normal state before it is performed, in the hope that an epidural anaesthetic and a

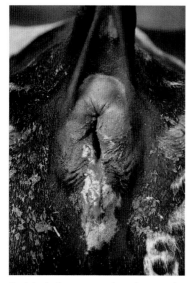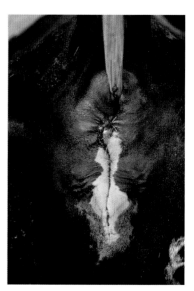

A third-degree perineal tear, the consequence of a traumatic calving, before, prepared for and after surgery.

large tampon placed into the rectum will allow the area to be adequately cleaned and prevent further faecal contamination of the area during surgery. It would also be hoped that a newly created surgical wound will heal better than the contaminated, damaged wound to the traumatised tissue that has resulted from the traumatic calving.

Tears to the Walls of the Vagina

Occasionally tears to the walls of the vagina may involve a vaginal artery. A bloody discharge from the vulva during and after calving is normal, but when this persists it should be explored, especially if there is a history of a traumatic calving. A cleaned and lubricated arm should be inserted gently through the vulval lips and into the vagina, taking the utmost care not to disrupt any blood clot that may have formed. If trauma can be identified and there is associated pulsatile bleeding from the area, urgent assistance needs to be requested and action needs to be taken. Until help arrives, if the bleeding end of a blood vessel can be

identified it should be pinched between finger and thumb to prevent further blood loss (despite the pins and needles and cramp this causes!).

If the source of the bleeding cannot be identified, for example if the torn end of the vaginal artery has retracted into the damaged tissue of the vaginal wall, an attempt should be made to apply pressure to the area to at least slow the rate of blood loss. This may be achievable by inserting a clean, rolled-up towel fully into the vagina and keeping it there until help arrives. A refinement may be to include a bag of frozen peas (sweetcorn will also be effective!) within the towel to promote local vasoconstriction.

Such cases always present a dilemma for the attending vet. They will hopefully arrive with minimal delay, but frequently by the time they do the bleeding might appear to have stopped. Removing the towel and investigating the situation risks starting it again, but this needs to be done because, although blood is no longer flowing from the vulval lips, the bleeding may not in fact have stopped, and the free blood is now running into the recently vacated uterus. With expert help and appropriate equipment it may now be possible to locate the bleeding end of the vessel and tie it off with suture material, or clamp it with artery clamps (which should be left in place for three or four days to be certain that bleeding will not re-start when they are removed), or even to suture the torn tissue around the bleeding vessel.

Tears in the Uterine Wall

Tears in the uterine wall often occur towards the tip of one of the uterine horns where the calf in distress during a difficult delivery has kicked a foot through the uterus, or dorsally close to and perhaps including the cervix where the dome of the calf's head has squeezed the uterine wall hard against the dorsal rim of the cow's pelvis, causing it to become bruised and friable. These tears can be difficult to identify, either because they are out of reach, or because of the continued presence of the 'cleansing' in the uterus. If they can be identified, surgery, either through the vagina or via the abdomen, may achieve a successful resolution; however, since the uterus following calving is not a sterile environment, peritonitis should be predicted and appropriate treatment given. (Small dorsal tears may heal satisfactorily without any action being taken as gravity will help prevent contamination of the abdomen, and uterine involution after the calving may close the defect – but this should not be relied on!)

Where tears to the uterine wall are not detected they should certainly be considered

A suckler cow just after calving with a prolapsed uterus. (Look at the size of the calf that has just been delivered!)

among the differentials where a cow fails to recover satisfactorily within the few days after calving. A typical history would be a call for veterinary assistance for a declining and sick cow classically four days after calving as peritonitis develops.

Prolapse of the Uterus

Perhaps the most dramatic possible consequence of a traumatic calving, although perhaps not the most serious (although it can certainly be severe), is a prolapse of the uterus following calving. Although these can occur after any calving, they are most common in young animals, particularly heifers, after a difficult delivery, and in older, fatter animals where there may also be an element of milk fever present. Needless to say, a prolapsed uterus is a medical emergency. The longer the organ remains prolapsed, the more likely it is to become traumatised, and if the weather is inclement, the more likely the affected animal is to suffer hypothermia as heat is lost from its wet, hairless surface.

To protect the organ from trauma it is important to keep the affected animal quiet and to prevent her from running around, assuming she can still stand. If she is recumbent, other animals should be moved away and a clean plastic sheet used to protect the uterus from contamination from the ground. If the placenta is still attached this should be left in place to provide additional protection against both trauma and heat loss.

Replacing the prolapsed organ is an art. An epidural anaesthetic should be given

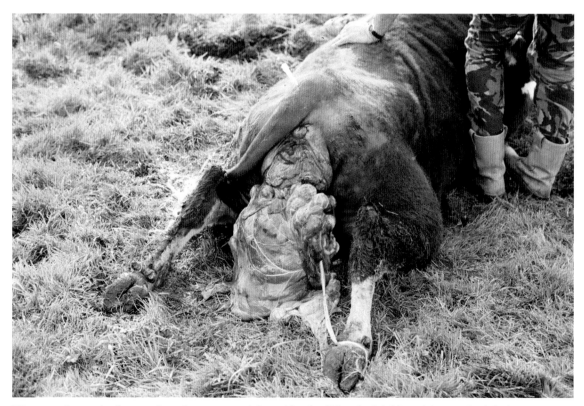

The cow shown in the previous image repositioned in the 'Plenderleith' position and with the prolapsed organ cleaned, but with the foetal membranes left in place, in preparation for replacement.

to prevent the cow 'pushing' as the uterus is replaced, and while this is taking effect the uterus should be cleaned of all straw and other debris, and then washed with plenty of warm water. If the affected animal remains standing, replacing the uterus can be facilitated by supporting the weight of the organ on either side of the animal at the level of the vulva. If the animal is recumbent, replacement of the organ can be made easier by positioning the cow in sternal recumbency with the hind legs pulled out behind her in a frog-legged or 'Plenderleith' position, and then kneeling with the bulk of the uterus in your lap as close behind the hind end of the cow as possible.

Replacement of the organ starts by massaging the part of the organ closest to the vulva back through the cervix and into the cow's abdomen, and then working backwards until it has all been repositioned. While doing this great care is needed to avoid puncturing the uterine wall with a finger or your hand; if pressure is needed it may be better to push with a closed fist than with an open hand.

Following replacement of the organ it is important to make sure that both uterine horns have been fully inverted (some people advise introducing a wine bottle, preferably empty, into each horn of the uterus to extend the reach and facilitate this); failure to do this risks the organ being prolapsed again. Suturing the vulva will not necessarily prevent this and should not be required if the uterus has been replaced and inverted effectively.

The same cow after replacement of the prolapsed uterus. In this case sutures have been used to close the vulva, probably to appease owner expectation rather than to retain the uterus in place.

An occasional, but particularly frustrating complication of a prolapsed uterus that occurs just after successful replacement of the organ is the death of the cow due to the release of blood clots that have formed in the stretched uterine vessels in the prolapsed organ: as the vessels return to a more normal diameter these clots are released into the circulation and lodge elsewhere in the body, perhaps in the lungs or the heart, blocking normal blood flow through these organs.

Some texts describe raising the hind end of the cow to assist in replacing the prolapsed organ, although the welfare aspects of this have been questioned; nevertheless the welfare aspect may be less severe than struggling to replace the organ for a prolonged period of time, both for the cow and for the veterinary surgeon!

Where replacement of the prolapsed tissue is proving difficult it is important to consider possible causes for this. Particularly challenging cases include the prolapse of both uterine horns, but with one inside the other; or if abdominal contents, including perhaps the bladder and small intestine, have also left the abdomen to sit within the prolapsed organ outside the cow.

Following replacement, additional medical treatment for the patient should be considered; perhaps oxytocin to promote uterine involution (although following calving and replacement of the uterus it is likely that the circulatory system is swimming with the hormone!), perhaps calcium to address sub-clinical or clinical milk fever, NSAIDs to make the cow more comfortable, and perhaps antibiotics if the organ has been heavily contaminated or damaged. Warm fluids containing calcium and an energy source for the cow to drink or provided by stomach tube will also improve the probability of a successful outcome.

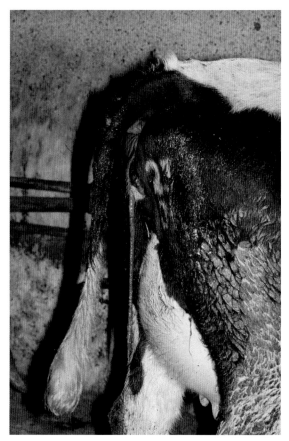

Retained foetal membranes.

Retained Foetal Membranes, Endometritis and Metritis

Perhaps the most common complication of parturition that might impact uterine health and future reproductive performance is a failure to pass the placenta. This would usually be expected within hours of the calf being delivered, but failure to achieve this is not uncommon following the delivery of twins, after a difficult birth, and where the dam suffers sub-clinical or clinical milk fever.

Usually retention of the foetal membranes (RFM) has little consequence for the general health of the cow, although infection resulting in endometritis or even metritis is inevitable and will delay uterine involution and extend the calving-to-conception interval. Treatment has been shown to be advantageous for future reproductive performance. Antiseptics and antibiotics have both been used locally within the uterus, but perhaps the most important aspect of any treatment is to return the cow to cyclicity and to induce oestrus using PGF if a corpus luteum is present on one or both ovaries.

Specific infectious agents that can impact uterine health are considered elsewhere within this text.

Cyclical Conditions Affecting Female Reproductive Performance

Reproductive success in the female depends on many factors, including satisfactory ovarian cyclicity and effective ovulation, which in turn depend on hormonal fluctuations and patterns, as discussed above. The age at which these begin, puberty, also depends on many factors, including genetic factors and breed, plane of nutrition and performance during the prepubertal period, and possibly even on the time of year during which each animal was born. Similarly, the time taken for a cow to undergo uterine involution and return to ovarian cyclicity after calving will depend on many of the same factors, particularly body condition and plane of nutrition; cows that are in good body condition at calving and fed a diet providing a high plane of nutrition will undergo uterine involution and return to ovarian cyclicity faster after calving than will cows that are in poor body condition at calving and fed a diet providing a poorer plane of nutrition. In both cases it may be possible to advance, if not cyclicity, then at least ovulation using exogenous progesterone. Although this may not be a sensible approach to the routine management of breeding, it might be useful to ensure that cows calving late during a fixed calving period at least get a chance to be served again before the subsequent serving period comes to an end.

Prolonged periods of progesterone dominance, as may be seen due to a failure of luteolysis, will inhibit ovulation and oestrus. This may be associated with uterine infections, pyometra, or the presence of a mummified foetus within the uterus. (The birth process is initiated by corticosteroid production by the near-term foetus. If, for whatever reason, the foetus dies *in utero* and is not aborted, it will not produce the necessary corticosteroids to initiate parturition. It may then remain in the uterus, preventing the cow from returning to cyclicity and conceiving again for months or even years.) The conditions described above will all have a direct adverse effect on future fertility, but the prolonged period of progesterone dominance over the female reproductive tract will also have a negative effect on future fertility even after successful treatment with PGF.

Cystic Ovarian Disease

Perhaps the most common problem affecting cyclicity in the cow, albeit that it is probably more common amongst dairy cows than beef cows, is cystic ovarian disease. This occurs when a dominant follicle fails to ovulate, perhaps as a result of abnormal hormonal patterns possibly resulting from an inadequate plane of nutrition (or metabolic stress in dairy cows); it then continues to increase in size. In many cases these cystic follicles will continue to secrete high levels of oestradiol, which may cause the affected cow to show persistent oestrus or even nymphomanical behaviour. Treatment in such cases might involve the administration of GnRH, aiming to lutenise the cyst, or the administration of exogenous progesterone. In some cases, however, the cyst will include luteal tissue and will have spontaneously become a lutenised follicular cyst. In such cases treatment with PGF may be indicated.

MANAGING MALE FERTILITY

If the consequences of an infertile cow in the herd are undesirable, the consequences of running an infertile bull can be nothing short of catastrophic; a single infertile cow means one fewer calf, but the bull, as they say, is half the herd, and without him in the peak of fitness and condition and fully fertile the consequence may be no calves! It is as important, therefore, in the suckler herd where reproductive management is based on natural service, to spend as much time trying to ensure male fertility as is spent on trying to ensure female fertility.

ASSESSMENT OF MALE FERTILITY

Spermatogenesis

Spermatogenesis is the formation of haploid sperm cells by mitotic and meiotic cell division within the seminiferous tubules in the testes, and is driven by the same hormones that drive ovarian function and the oestrus cycle in the female. Gonadotrophin releasing hormone (GnRH) from the hypothalamus in the brain acts on the anterior pituitary gland, stimulating the release of luteinising hormone (LH) and follicle-stimulating hormone (FSH) in a pulsatile manner. LH, in turn, stimulates the release of testosterone from the Leydig cells, which are found adjacent to the seminiferous tubules in the testes, and this, along with FSH, stimulates cell division of the germinal epithelium in the basal layer of

the seminiferous tubules. The resulting cells, supported and nurtured by the Sertoli cells, develop from primary and then secondary

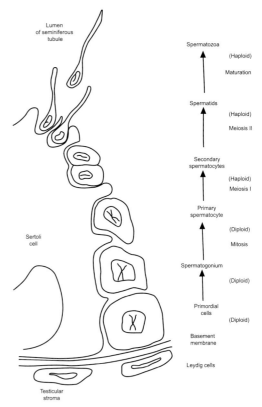

Schematic representation of haploid sperm cell production from diploid primordial cells, spermatogonia and primary spermatocytes through haploid secondary spermatocytes and spermatids, which then mature into spermatozoa, by first mitotic and then meiotic cell division, all supported by the Sertoli cells in the seminiferous tubules in the testes.

spermatocytes to spermatids, and then spermatozoa: these need to undergo a further period of maturation as they travel through the epididymis, with the whole process taking about two months, before becoming fully fertile.

There is much that can go wrong during the process of spermatogenesis! However, the production of high quality, fully fertile spermatozoa is only part of what needs to be considered in the assessment of male fertility; even if a perfect sample of semen is produced, if the bull cannot detect when a cow is in oestrus, get to her, mount her, achieve intromission and deliver his semen into her anterior vagina, results will be disappointing (or non-existent!). There is much more to male fertility than semen!

Physical Assessment

The assessment of male fertility should begin, as should the examination of any animal in a veterinary context, from a distance, and then with a full clinical examination, along with an assessment of expectation. Is the animal old enough, remembering that different breeds reach puberty and maturity at different ages, and how many females is he expected to impregnate? (While an experienced, mature, fully fertile bull can be expected to successfully serve fifty cows over a period of ten weeks, a useful rule of thumb for immature bulls is to run them with the same number of randomly cycling females as their age in months, once they have reached puberty.) Is he sufficiently well grown and in a 'fit but not fat' body condition? (An emaciated bull is unlikely to be producing good quality sperm, or any sperm at all, while an obese animal with an excessive layer of fat surrounding the testes, which will affect thermoregulation, is also likely to produce sperm of compromised quality.)

Are there signs of conformational abnormality or extreme (particularly with respect to foot and leg conformation), or of pathological processes that are, or could be expected to affect mobility and mounting ability? Are there any signs of infectious or contagious disease, remembering that some will show no clinical signs either for a prolonged period after infection – Johne's disease for example – or ever – Campylobacter foetus venerealis – and that some may be congenital; could he perhaps be persistently infected with BVD (Bovine Viral Diarrhoea)?

Assessing the External Genitalia

Following a satisfactory physical examination of the animal, attention can be turned to his reproductive apparatus, to include an assessment of both the external apparatus – the penis, testicles and epididymis – and the internal accessory sex glands – the ampullae, seminal vesicles and prostate gland (and bulbourethral glands), which all contribute to the ejaculate.

The sheath, and particularly the preputial orifice, should be inspected for abnormalities including, but not limited to, warts, evidence of previous trauma, stenosis of the orifice, ulceration or discharge. With great care the penis should then be palpated within the sheath to ensure the absence of adhesions or any swelling that might suggest a previous rupture of the corpus cavernosum penis. In young bulls it may also be possible to palpate the sigmoid flexure just proximal and caudal to the neck of the scrotum.

The neck of the scrotum should be inspected to confirm the absence of herniation of any abdominal content through the inguinal ring. Scrotal circumference, which correlates well with fertility, should be measured: the minimum requirement for a mature bull to pass a breeding soundness examination is recommended to be 34cm, although this can be progressively reduced to

Male reproductive anatomy

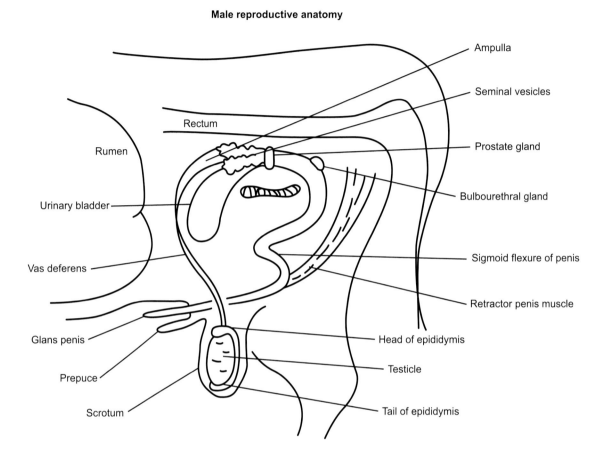

Ampulla

Seminal vesicles

Rectum

Rumen

Prostate gland

Urinary bladder

Bulbourethral gland

Vas deferens

Sigmoid flexure of penis

Retractor penis muscle

Glans penis

Head of epididymis

Prepuce

Testicle

Scrotum

Tail of epididymis

Male reproductive anatomy.

30cm in bulls of a year of age. The testicles should be palpated within the scrotum, again with great care, ensuring that they are of uniform size (but not necessarily exactly the same!) and consistency: firm but not hard (often described as feeling like a ripe melon, a flexed bicep or a new tennis ball – take your pick!). The tail of the epididymes should also be palpated to confirm normality, with the body and head of the epididymes also being palpable in all but the biggest animals.

Assessment of the Accessory Sex Glands

After assessing the external genitalia, a gloved and lubricated hand should be gently inserted into the rectum and an assessment made of the accessory sex glands: the seminal vesicles should be palpable as soft, lobulated (like small bunches of grapes) masses on either side of the penis, and the prostate gland as a firm band of tissue crossing the body of the penis caudally. Gentle palpation of these organs should not be resented, and massage will frequently result in the testicles being pulled up within the scrotum (confirming the absence of adhesions) and the glans or tip of the penis being exteriorised, which allows visualisation to confirm normality. It can also facilitate the collection of an ejaculate of semen (but if this is the intention, beware of causing ejaculation before being prepared!).

SEMEN COLLECTION AND EVALUATION

Now is the time to move on to semen collection and evaluation, if required. The gold standard method of collection is to use an artificial vagina (AV), but this requires the bull to be relatively amenable to handling, a suitable animal in oestrus for the bull to mount, patience (sometimes!) and split-second timing. It remains the method of choice when semen is being collected for the preparation of straws for artificial insemination (AI), but with the development of improved equipment, electroejaculation (EEJ) has largely replaced the use of the AV for on-farm semen collection and evaluation because it can be carried out in a more controlled, and therefore safer, environment, collection can almost always be predictably achieved, and the ejaculate produced usually correlates well with what would have been produced using an AV. (If all else fails, semen can be retrieved from the vagina of a cow after service, but in this case the evaluation of the sample will be compromised.)

Electroejaculation involves inserting a probe gently into the rectum of an adequately restrained bull after massaging the accessory sex glands, and then gradually increasing the current passing between the electrodes, maintaining due regard for the welfare of the bull, in a pulsatile manner until ejaculation is achieved. It is vital that the equipment used to collect the sample and then to examine it is clean, dry and warm.

The ejaculate is assessed by eye for volume and appearance (creamy, milky or watery, which correlates with sperm density), then undiluted under low power microscopy to assess gross motility (wave motion), which is scored between 0 and 5 (0 - no movement at all, 1 - no swirl but a generalised oscillation of individual sperm cells, 2 - a very slow but

The author with a semen sample he has just collected by electroejaculation.

distinct swirl, 3 - a slow, distinct swirl, 4 - a moderate to fast swirl with dark waves, 5 - a fast, distinct swirl with continuous dark waves), and then diluted with phosphate-buffered saline under higher power microscopy to assess the progressive motility of the individual sperm cells. A minimum of 60 per cent of the sperm cells present are required to demonstrate progressive motility for a bull to pass a breeding soundness examination. If this is not achieved this may be due to what is termed a 'rusty load': sperm cells that have spent a prolonged period of time within the epididymes during a period in which the bull has been inactive, and which have become senescent. In such cases a second ejaculate should be collected for examination.

Examining Individual Sperm Cells

The final step of a breeding soundness examination involves examining individual sperm cells under high power microscopy to assess morphology: this is done using a stained (usually eosin/nigrosin) smear prepared from the ejaculate, or using phase-contrast microscopy to examine sperm cells that have been preserved in formol saline. This is a highly skilled task involving the examination of a minimum of one hundred individual spermatozoa. (Although automated processes are now available, they have not yet totally replaced the role of an experienced clinician.) For a bull to pass a breeding soundness examination a minimum of 70 per cent of the sperm cells present in the ejaculate must be deemed morphologically normal, with no more than 20 per cent showing nuclear defects.

Sperm cell abnormalities can be classified in a variety of ways, including based on the region of the sperm cell that is affected (head, mid-piece, tail), according to their effect on fertility (major, including most abnormalities of the head and mid-piece, and minor, including bent tails and retained distal cytoplasmic droplets), and the stage at which the abnormality is triggered. Primary abnormalities, for example pyriform heads, nuclear vacuoles, diadem defects and knobbed acrosomes, occur during spermatogenesis; secondary abnormalities, including swollen acrosomes, detached heads, distal midpiece reflex and retained cytoplasmic droplets, occur during sperm cell maturation within the epididymis; and tertiary defects, most commonly bent tails, occur during collection, usually as a result of either cold or osmotic shock. As a rule of thumb, primary abnormalities are more significant than secondary abnormalities, which in turn are more significant that tertiary abnormalities, and the greater the

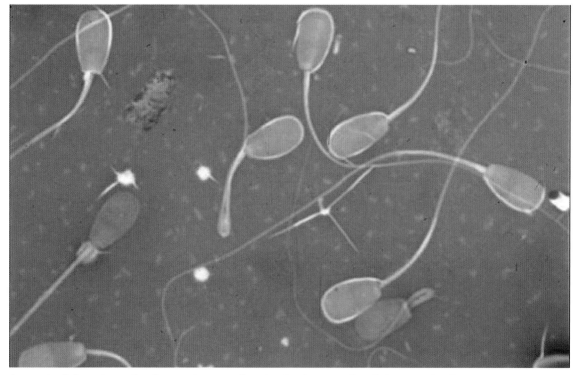

An eosin-/nigrosin-stained spermiogram showing both normal and abnormal cells.

proportion of spermatozoa with any one abnormality, remembering that some have a genetic basis and are heritable, the more significant it is likely to be.

INFECTIOUS DISEASE AND BULL BIOSECURITY

Many of the infectious diseases that can affect female fertility can also affect male fertility; the more significant of these are discussed elsewhere in this publication. The bull, however, arguably deserves special consideration with respect to infectious disease and biosecurity because in herds where natural service is the major or only reproductive strategy the bull will be in close (intimate!) contact with every (hopefully!) breeding female during the serving period – and what better way could be designed for spreading an infectious agent as quickly and as widely as possible within a herd of cows?! Also many herds that are described as 'closed' (that is, which never introduce, or re-introduce, an animal that has been born or is resident in another herd) will purchase and introduce the occasional breeding bull. Whenever this occurs careful consideration should be given to the health status of the bull, the possibility that he might be introducing more than just the desired genetics to your herd, and what steps can be taken to ensure that this is not the case.

Sourcing a new bull should be considered well in advance of when he will first be required to work, to allow time for relevant investigations to confirm his health status, and any required preventative medicine protocols, vaccination for example, to be carried out while he remains in quarantine, away from the herd he will, hopefully, eventually be added to.

It is important to confirm the health status of the bull's herd of origin (and, if possible, his natal herd and other herds he may have been resident in, if these are not the same) prior to purchase. In an ideal situation the herd will have an accredited health status for some of the more relevant diseases under a CHeCS (Cattle Health Certification Standards) scheme. These schemes work to a set of rules that define required biosecurity precautions that need to be taken, not only when adding animals to the herd and at farm boundaries, but which are also relevant to the transport of animals, showing animals, and visitors to the farm, and the health screening required – including which animals to sample, how many and when – to provide reasonable reassurance of the absence of infection. CHeCS schemes exist for BVD (Bovine Viral Diarrhoea), IBR (Infectious Bovine Rhinotracheitis), bTB (Bovine Tuberculosis), Johne's disease, leptospirosis and neospora.

In the absence of a CHeCS accredited herd health status, evidence of an active surveillance policy should be examined and also the results of this surveillance, and the biosecurity precautions taken should be carefully considered, particularly when animals are added to the herd: the more animals that are added to the herd, the more frequently they are added, and the greater the number of herds they are sourced from, the greater the risk of the presence of infectious disease within the herd. It is not enough just to be told that all calves born are screened for BVD virus using tag-and-test technology – you also need to know whether any PI ('persistently infected'; *see* 'BVD' elsewhere in this book) calves have ever been identified and when this was. In the case of bTB it would be illegal for the bull to be sold without his herd of origin having OTF (officially bTB-free) status, and the bull himself having passed a pre-movement 'skin' test (if required). Furthermore, how long the herd of origin has had a continuous OTF status remains highly relevant, and also when the most recent skin-test reactor was removed from the farm.

Once the bull has been sourced and placed in quarantine, a proposed (but not exhaustive) preventative medicine programme might include testing for the following:

BVD

A blood sample is collected for testing for both virus and antibody. In most cases, if an animal is antibody positive for BVD this is usually interpreted as meaning that it cannot be persistently infected (PI) with the virus – but this is not the case with certainty. Young PI animals, particularly if the offspring of a non-PI dam, may test antibody positive due to residual maternal colostral antibodies (which can be predicted to be present at high levels thanks to the continuous boosting effect of the *in utero* PI calf on the maternal immune system), which may persist for some months. Also animals that are persistently infected with, for example, a type I strain of the virus can produce low levels of antibodies if they are challenged with a different, type II strain of the virus. Even when a bull is antibody positive and virus negative, consideration should be given to the need for testing the animal's semen for the continued presence of the virus, even though persisting testicular infection is extremely rare.

Vaccination should be carried out, despite the fact that many BVD vaccines are not licensed for use in breeding bulls. The entire primary course of vaccination (although this may involve only a single dose of vaccine) should be completed at least two weeks, and preferably a month, before the bull is released from quarantine and introduced to the herd.

IBR (BoHV1)

A blood sample is collected for testing for antibodies to the virus. If the bull has not been previously vaccinated against IBR a gB

test should be used. If the bull has previously been vaccinated a gE test should be used to differentiate between exposure to field-strain virus, and antibodies raised as a result of vaccination with a 'marker' vaccine. (There is currently no way of differentiating antibodies raised as a result of exposure to field-strain virus from antibodies raised as a result of vaccination with a 'conventional' vaccine, which might have been given many months or even years ago as part of a multi-valent pneumonia vaccination strategy.) A positive antibody result to the appropriate test should be interpreted to mean that it is highly likely that the bull is latently infected with BoHV 1, the cause of IBR, and if maintaining an IBR-free herd is important the bull should not be used.

Vaccination should be completed, if required, before the bull is released from quarantine.

bTB

As well as any statutory testing that is required, a post-movement skin test between 60 and 120 days after the arrival of the bull, and the use of one or more of the newer, more sensitive testing technologies that are becoming available, should be considered.

Johne's Disease

The sensitivity of the current serological tests used to detect antibodies to Map (*Mycobacterium avium* subspecies *paratuberculosis*), the cause of Johne's disease, is generally agreed to be poor, particularly in young animals early after infection. It may be possible to improve test sensitivity by testing an 'anamnestic' sample collected between five and 30 days after the administration of the PPD used when TB testing. In addition, faecal testing can also be used to try to improve diagnostic sensitivity, but this requires that Map is being shed in the bull's faeces, which can be intermittent and does not start immediately after infection. Furthermore the different PCR tests that are now in more common use than either solid agar or liquid culture may detect different populations of infected animals, so perhaps more than one PCR test should be used and on multiple occasions.

Leptospirosis

A blood sample is collected for testing for antibodies to the bacterium. A negative result is reassuring, but a positive result requires careful consideration. There is currently no way of differentiating between antibodies raised as a consequence of infection and which may, therefore, indicate a carrier status, and antibodies raised as a consequence of previous vaccination. Antibiotic treatment may be successful at eliminating infection and the carrier state, but this cannot be guaranteed, and what to use and for how long? Many texts suggest that the administration of a single double dose of Dihydro-streptomycin should be sufficient to eliminate infection, but this advice has persisted from the historic situation when the use of such a product in dairy cattle did not require milk produced by treated animals to be withdrawn from sale for human consumption. Many other antibiotics are equally as effective as Dihydro-streptomycin in the elimination of leptospirosis.

An entire primary course of vaccination, if required, should be completed before the bull is released from quarantine and added to the herd.

Campylobacter

If the bull is a virgin, venereal campylobacter infection can be discounted – but is he really a virgin? Many bulls sold as virgins have 'just been run with one or two cows to make sure they can work'. This is not a virgin!

If the bull is not a virgin, testing to detect infection can be considered, but this is not straightforward, and there is no blood test

currently available that will define status. The best that can be done is to take aseptic vigorous sheath washings, concentrating particularly on the area of the fornix, using phosphate-buffered saline, with the aim of dislodging the bacteria from their position deep within the crypts and crevasses within the prepuce. After 'washing' the interior of the prepuce the phosphate-buffered saline is added, some after being filtered, to antibiotic-containing culture media (to inhibit the growth of non-target organisms); this is then submitted as soon as is possible – and certainly within twenty-four hours of being collected – to a diagnostic laboratory for culture and fluorescent antibody testing.

A positive result indicates that the bull is, indeed, infected – but how to interpret a negative result? Is the bull uninfected, is it infected but there was a failure to recover the infectious organism in the preputial wash, or was the infectious organism recovered in the preputial wash but became non-viable before reaching the laboratory? This dilemma leads many to bypass sheath washing for diagnostic purposes, and to carry out prophylactic sheath washing with an antibiotic preparation often made by mixing the contents of a 100ml bottle of penicillin and streptomycin with 200ml of food-grade arachis (peanut) oil to make it sticky and prolong its presence within the prepuce.

After cleaning the preputial orifice and trimming any dangling hair, approximately half of the mixture, along with 20ml of air, is infused into the prepuce in the region of the fornix using a firm but flexible catheter; this is then withdrawn whilst holding the preputial orifice closed with one hand to prevent the immediate loss of the washing mixture. Using the other hand, the prepuce is then vigorously massaged 'one hundred times' (this is hard work!) to distribute the washing mixture into all the folds, crypts and crevasses within the prepuce, particularly in the region around the fornix.

Having completed this procedure the mixture is allowed to drain from the prepuce; then after a short interval, the process is repeated using the remaining half of the washing mixture. This entire process is then repeated daily on the following two days. But even after this, can you be certain that the infection, if present, has been eliminated?

SPECIFIC PROBLEMS AND ABNORMALITIES AFFECTING THE MALE REPRODUCTIVE TRACT

Although there are many and varied problems and abnormalities that may affect one or more parts of the male reproductive tract, some resolvable, others irresolvable but all of which retain the potential for an adverse effect on reproductive performance, only those more common abnormalities, and therefore those more likely to be encountered, will be discussed here.

Problems and Abnormalities Affecting the Prepuce and Penis

Although congenital problems affecting the prepuce and penis are seen – stenosis of the preputial orifice preventing extrusion of the penis, and the presence of a persistent frenulum, for example – acquired problems are much more common in practice and so are likely to be more significant. (A persistent frenulum is the usually membranous residue at the tip of the penis, of an attachment that exists in the developing foetus connecting the penis

A small but neglected preputial prolapse that would probably have been easy to resolve when fresh, but which is now fibrosed and will require surgical resection if the bull is to continue as a breeding animal.

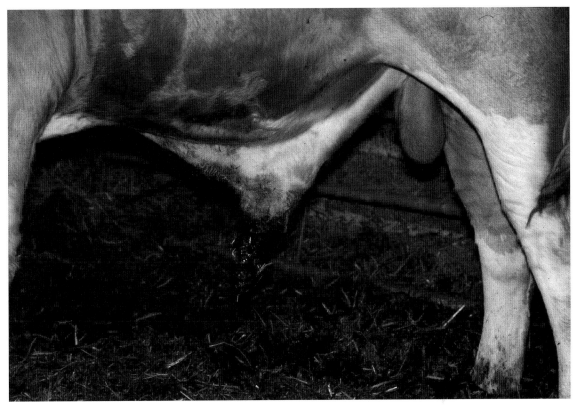

A larger, fresh preputial prolapse that it may still be possible to reduce.

to the prepuce, but which usually breaks down spontaneously. If it does not, it will cause a deviation of the penis preventing intromission. It can often be remedied by simply sectioning the remnant, although this may be quite fleshy and may bleed.)

The most commonly encountered acquired preputial problems are likely to be prolapses, possibly with associated trauma and non-specific infection. (Infectious pustular balanoposthitis, caused by bovine herpes virus 1, and *Campylobacter foetus venerealis*, are specific infections that are discussed elsewhere in this book.) While preputial prolapses are not necessarily uncommon, or even a problem if the prolapsed tissue can be spontaneously withdrawn back into the sheath, anecdote suggests that they are more likely to be significant and become traumatised in bulls where the prepuce

dangles below the ventral abdominal wall, rather than in bulls where the prepuce is tight to the abdominal wall – this is particularly the case where the animals are grazing areas of rough scrub, or the bull has a tendency to disregard fences, particularly barbed-wire fences.

While preputial prolapses will rarely be life-threatening (assuming the bull can still pass urine), whether traumatised and infected or not (preputial infection often responds well to lavage, cleansing and topical treatment, and traumatic damage usually heals well even in severe cases), they can have severe consequences for breeding potential. Even relatively minor prolapses, if they persist, will become inflamed and fibrotic, leading to occlusion of the preputial orifice, eventually resulting in an inability to serve. Small, persisting prolapses may therefore need to be replaced and retained within the prepuce with a purse-string suture around the orifice to allow resolution if breeding potential is to be retained. Larger prolapses may require surgical amputation, or may even be deemed hopeless.

A wide range of acquired penile issues will also be encountered in practice, including, for example, prolapse (perhaps as a consequence of damage to the retractor penis muscle, or as an occasional complication of sedation), trauma and infection, penile warts (which it may be possible to pinch off but are likely to bleed profusely and will usually regress in time – but it may be that the bull is needed for use now!) or neoplasia (penile squamous cell carcinomas are recognised, occasional problems that may be amenable to surgery although this is likely to be rejected on economic grounds in the majority of cases), and even encircling foreign bodies, hairbands for example. Perhaps the two most commonly encountered, and therefore arguably the most significant, acquired penile problems affecting breeding bulls are deviations and rupture of the *corpus*

cavernosum, commonly referred to as a 'broken willy'.

Deviations, usually sideways or ventrally and classically spiral, where the tip of the penis corkscrews as erection is achieved, occurs when the dorsal ligament of the penis slips from its position dorsally on the penis, pulling the tip sideways, down or round as the bull attempts to serve. In the mildest of cases, where intromission remains possible, this need not have an immediate adverse consequence for the breeding potential of the bull, but most cases are progressive, and at the point when the deviation prevents successful intromission, the bull obviously has no future as a breeding animal. Surgery to secure the dorsal ligament of the penis in its normal anatomical position is described, but the ethics of attempting to correct a condition that is, at least in part, inherited, are questionable, and, as with other penile surgery, usually rejected on economic grounds.

Rupture of the *corpus cavernosum* is often indicative of poor bull management – young, small bulls being run with large, mature

A four-year old Charolais bull with a history of good fertility last year, but which this year is failing to successfully impregnate cows due to a spiral deviation of the penis.

cows too tall for them to reach are commonly affected. Bulls have a fibro-elastic penis, with full erection achieved by the engorgement of vascular chambers, the largest of which are the two *corpora cavernosa*, with blood under considerable pressure. If, during the ejaculatory thrust, the tip of the penis fails to enter the female reproductive tract and instead slams into the hind end of the cow, the wall of the *corpus cavernosum* can split, leading to bleeding into the surrounding tissues and the formation of a haematoma: this presents as a swelling, usually up to the size of a large orange or a small grapefruit, encircling the shaft of the penis, usually cranial to, but sometime caudal to, the scrotum. There is usually an associated small prolapse of the tip of the penis.

The consequence of such damage is that the bull will be unable to achieve an erection and intromission, and so will no longer be able to work as a breeding bull. (The condition is not, however, fatal, or even necessarily a welfare problem, so beware when insuring breeding bulls, because rupture of the *corpus cavernosum* will only be covered under a 'loss of use' policy, and not if the bull is only insured for death!) The situation is not, however, totally without hope. In time the damage to the wall of the *corpus cavernosum* will heal, although a residual weakness should be predicted, and small haematomas may organise and resolve spontaneously. Where a larger haematoma forms, surgery is described to evacuate the organising blood

Rupture of the *corpus cavernosum*. Note the swelling around the penis cranial to the scrotum, and the prolapse of the tip of the penis.

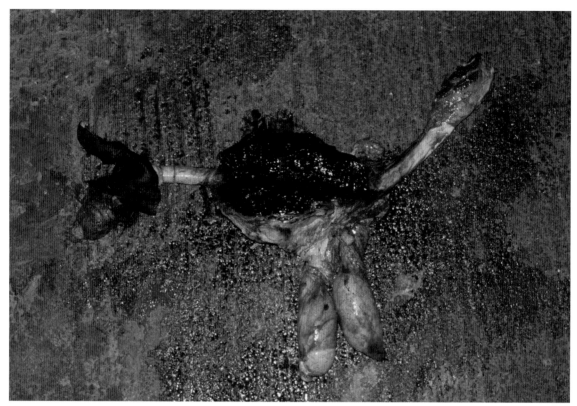

A dissection of the male reproductive tract from a bull that has suffered a rupture of the *corpus cavernosum*, showing the extent of the associated haematoma.

clot; however, this should not be carried out, if it is attempted at all, until it is certain that the bleeding has stopped and that the damage to the wall of the *corpus cavernosum* is healing.

Following surgery, a significant period of recovery will be necessary, during which gentle teasing of the bull will be required to prevent the formation of adhesions within the prepuce (or to break down any that have formed), which may prevent the bull exteriorising the penis to serve cows in the future, even if the procedure is successful. Given the cost of this, not only in monetary terms but also in terms of time and management input, with a somewhat unpredictable outcome and a propensity for reoccurrence even if a successful outcome

is achieved, it is often rejected as a viable treatment option in common with many other penile surgeries possible.

One further acquired problem affecting the male reproductive tract that should perhaps be mentioned for completeness is urolithiasis, although this is more a problem that affects animals being intensively fattened and is rarely seen in breeding bulls; it is therefore discussed elsewhere in this book.

Problems and Abnormalities Affecting the Scrotum and Testes

Essentially these problems involve number, size and consistency: is the scrotum enlarged, are two testicles present and fully descended into the scrotum, and are they too big or too

Herniation of abdominal content into the neck of the scrotum.

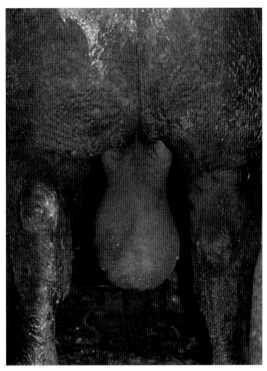

Enlargement of the right testicle.

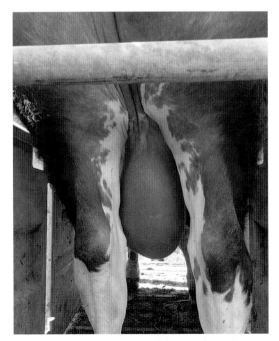

Generalised enlargement of the scrotum due to haematoma formation following a kick.

small, too soft or too firm, or do they contain 'lumpy-bumpy' bits?

Enlargement of the scrotum, assuming the presence of normal testicles, will be due to a relatively limited number of possible causes. Where the scrotal neck is not involved the most likely causes are blood, either due to local trauma and haematoma formation or to the presence of an abnormal vasculature (termed a varicocele), or pus, again possibly due to local trauma or haematogenous bacterial spread, and abscessation. Where the neck of the scrotum is also swollen, inguinal herniation of abdominal content, usually limited to part of the mesentery but possibly including loops of small intestine, should be considered.

Although in theory each of these conditions is treatable, the value of the bull and the prospects for returning him to full fertility should be considered: if this is not achieved, the consequences need to be carefully considered before embarking on complex and expensive procedures and programmes of medication. In most cases, given the uncertainty of success, these will not be warranted, and the affected animal will be culled and replaced with another, problem-free bull.

Changes in testicular size and consistency, affecting either one (unilateral) or both (bilateral) testicles may be congenital, for example in cases of congenital hypoplasia, or acquired, usually as a consequence of trauma or infection. Again, treatment may be possible and may successfully address an infectious agent that may be present, but residual fibrosis within the testes is likely to compromise future fertility, causing the breeding potential of the animal to be uncertain. Again, unless the animal is particularly valuable, replacement is the usual course of action.

Occasionally cryptorchid 'breeding bulls' will be encountered with a single, apparently normal testicle fully descended within the scrotum. Careful palpation may identify the second testicle partly descended within the inguinal region, but it may have been retained within the abdomen and remain undetectable. Such bulls will be able to successfully impregnate cows, but capacity will be reduced and they would not usually be regarded as being suitable for breeding.

Problems and Abnormalities Affecting the Accessory Sex Glands

Problems involving the accessory sex glands will usually not be as obvious as some of those that might affect other parts of the male reproductive tract. Some indication of their possible presence may, however, be gained by the careful examination of semen samples: the presence of pus or an increased number of white blood cells

in the sample, for example, may raise suspicions. Confirmation will, however, require palpation and possibly ultrasound examination of the glands *per rectum* by an experienced veterinary surgeon.

Perhaps the most likely issue to be encountered involving the accessory sex glands is seminal vesiculitis, an inflammation of the seminal vesicles. In the acute phase this will result in an enlargement of the glands, which will then often be painful on manipulation. In more chronic cases the size of the glands may be more normal and a pain response on manipulation reduced or absent, but the glands will frequently feel firmer than they normally would.

A wide range of potential infectious causes for seminal vesiculitis has been described. Cases seem most prevalent in young bulls, and particularly those being fed energy-dense diets. The success or otherwise of antibiotic treatment is reported as having mixed results (and spontaneous recovery can and does occur, particularly in young bulls), with Macrolides generally being considered the antibiotic selection of choice. Treatment early during the course of the disease is considered to carry the best probability of a positive outcome.

CASTRATING BULL CALVES

Castrating bull calves is, in most cases, a routine husbandry procedure aimed at facilitating handling, and removing the risk of unwanted pregnancies. Despite this, it remains significant surgery controlled by statute and with the potential for complications, particularly (depending on the method chosen) haemorrhage and infection, and consequent compromise to welfare. The requirement for the procedure and the method employed should therefore be carefully considered each time it is carried out.

In the UK castration can be effected by the application of a rubber or 'elastrator' ring around the neck of the scrotum above both testicles (this should be checked carefully before final placement of the ring): this prevents blood flow to the testicles and scrotum, causing the tissue below the ring to die, wither and slough; however, this method is only permitted to be used to castrate calves under a week of age. As a method, it has the advantage of eliminating any concerns about bleeding, and almost eliminating the possibility of infection (although occasionally local infection around the ring occurs, and in very rare cases tetanus may be seen following 'ringing').

There is a possibility, however, that if the ring is placed too high up the scrotal neck the urethra may be included within it, preventing the calf from urinating and leading to almost inevitable death in the most distressing of circumstances. Even when the ring has been placed without problem there is a growing expectation that calves castrated in this manner should receive a dose of non-steroidal anti-inflammatory (NSAID) medication at the time the ring is applied, and the routine use of a local anaesthetic infused into the neck of the scrotum before the ring is applied is also debated.

Despite the advantages of 'rubber ringing' during the first week of life, many farmers prefer to leave their bull calves entire for a variable period after birth, giving as their reason that entire bull calves perform better. However, this has to be set against the greater risk of later castration, both to the calf and the operator, and in the UK, the legal situation; local anaesthetic is a legal requirement (and the administration of a NSAID would be expected), and only a veterinary surgeon may castrate calves over two months of age. (This situation may differ in other countries.)

Two main methods exist for the castration of older calves in the UK: using a 'burdizzo' or 'bloodless castrator', or surgical castration. Burdizzos are veterinary instruments designed to crush the spermatic cord and permanently disrupt the blood supply to the testicles, causing them to atrophy and cease to function. After administering local anaesthetic and NSAIDs the first testicle is drawn down into the scrotum. The burdizzo is placed with its jaws across the neck of the scrotum just far enough to include the spermatic cord, and the handles are closed to crush the cord. The instrument should be left in place for thirty seconds or thereabouts to ensure disruption of the blood supply to the testicle. The procedure should be repeated,

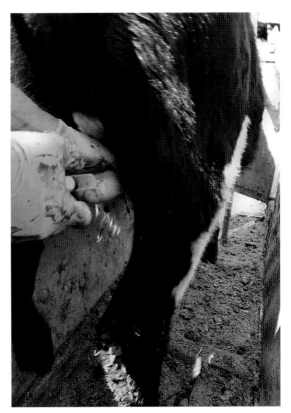

Local anaesthetic being administered prior to surgical castration.

first on the same side but at a slightly different position on the scrotal neck, and then twice on the other side of the scrotal neck to crush the contralateral spermatic cord.

Although castrating calves with burdizzos has the significant advantages of eliminating the risk of haemorrhage and almost eliminating the risk of infection, success does depend on the skill of the operator; furthermore the degree of testicular swelling and the discomfort experienced by the calves on which it is used, which may persist for some days, makes many involved in rearing calves uncomfortable with this method of castration.

Surgical castration has the advantage of knowing that when two testicles have been removed the job has been done. It does, however, present a greater risk of both infection and haemorrhage.

With the best will in the world, the task is unlikely to be carried out within a sterile operating theatre, or even under conditions even remotely approaching sterility, although cleanliness is important. Debate continues as to the value of washing and disinfecting the scrotum prior to surgery, particularly if it is visibly clean and dry (and if it is not, then perhaps the procedure should be left for another day!), and whether the routine administration of long-acting antibiotics is sensible to try to prevent infection. (Tetanus is an occasional recognised complication of open surgical castration in farm animals, so while it is not common, it might make sense to ensure all animals presented for surgery have been vaccinated to protect against this possibility before the surgery is carried out.) What is important, however, is that the procedure should be carried out in as clean and dry an environment as possible.

Haemorrhage remains a risk of surgical castration, irrespective of the method employed, and the larger the animal being castrated the greater the risk. In younger animals, the elastic recoil and retraction of the blood vessels in the spermatic cord after twisting and a 'long, controlled' pull may be sufficient to control bleeding and prevent haemorrhage from being an issue. If bleeding does persist, the common solution of tying a length of baler twine tightly around the neck of the scrotum will not stop it, as the bleeding end of the spermatic cord will have retracted into an inaccessible position within the abdomen. (What it will do, however, is stop the blood pouring from the scrotal incision, providing some reassurance to the owner while the surgeon prays for the blood to clot!)

In larger animals an emasculator might be used, which, when correctly applied (nut to nut!), will sever the spermatic cord and the associated blood vessels while applying pressure to 'crimp' the blood vessels above the cut to promote haemostasis. The emasculator should be held in place for sufficient time for the blood to at least start clotting, and so has the disadvantage of prolonging the procedure, whilst also increasing the risk of infection. When castrating even larger animals, it is often wise to place a transfixing ligature in the cord above the site at which it is sectioned, to ensure haemostasis.

One further potential complication of surgical castration, especially when the 'cut and pull' method is used, is 'gut tie'. As the severed and bleeding (at least to some extent) end of the spermatic cord retracts into the abdomen under tension, its final resting place cannot be predicted with any certainty: it may come to rest alongside the bowel, or even wrap around the bowel. The blood clot that forms at the severed end of the spermatic cord, before organising, forming scar tissue and retracting, may, therefore, on occasion include a segment of gut, which may become squeezed or even occluded some considerable time after the surgery to castrate the animal, presenting as colic and, in the most severe cases, an inability to pass faeces. It may be possible to perform further surgery to break down the adhesions and restore patency to the gut in such cases, but emergency slaughter is more likely to be the outcome.

Assuming a clean, dry environment and adequate facilities to safely restrain the animals to be castrated, the procedure can continue. Local anaesthetic (procaine is the licensed option in the UK) is injected either into the neck of the scrotum on each side, or into each testicle and then under the skin of the scrotum as the needle is withdrawn. Whilst waiting for the anaesthetic to take effect, other medication can be given: a dose of a NSAID and antibiotics, and fly control if considered necessary. The scrotal incision is a matter for personal preference: some surgeons incise the scrotal skin caudally on both sides along the long axis of each testicle (whether to cut through the testicular tunic or not, to perform an 'open' or 'closed' castration is also a matter for personal preference, although a better grasp of the testicle is possible if the tunic is not incised), others cut off the bottom of the scrotum completely. The testicles can then be exteriorised and removed one at a time, either by twisting and pulling, or using emasculators as described above.

After removal of the testicles the remaining scrotal skin should be pulled downwards to ensure no tissue remains protruding from the wound(s), which could act to wick infection into the surgical site; if protruding tissue does remain it should be removed. The scrotal incision(s) are then usually left open to allow drainage, but are coated with a topical antiseptic before the animal leaves the crush. They would be expected to dry up and scab over within a couple of days, after which time the animal can be treated as normal.

Cryptorchid Animal and Rigs

Cryptorchid animals are those where only one testicle has descended into the scrotum. Rigs are those where a rubber ring has been inexpertly applied soon after birth so that the scrotal blood supply has been occluded and it has sloughed, but one or both testicles have been retained in an inguinal position above the ring. These animals present a particular challenge when castrating calves, not least about whether surgery, in the case of congenitally cryptorchid animals, should be carried out. Although retained testicles are unlikely to produce viable sperm because of their higher than optimal temperature, they will still produce male hormones and so the animal will still act like a bull as it matures; furthermore if one normally descended testicle is present within the scrotum the animal is almost certain to remain fertile.

If both testicles can be identified, albeit in an abnormal position, surgical castration should be possible (although it might be considerably more challenging than castrating a similar animal with both testicles normally positioned within the scrotum). If only one of the testicles can be identified, then it might be preferable not to proceed with the surgery but to leave the animal entire, remembering to treat it as a bull for the remainder of its life.

CHAPTER 6

CALVING THE COW AND CARE OF THE NEW-BORN CALF

Having successfully managed the cow to the point of calving at the desired time and in optimum body condition, her entire productivity for the year now depends on the successful delivery of a live calf: it all comes down to this! Of course, with luck, influenced by expert management, the aim is for each cow to proceed through the stages of labour without need for assistance: stage one, during which the cervix dilates to allow passage of the calf; stage two, during which powerful contractions of the muscular uterine wall expel the calf; and stage three, which is completed when the placenta or afterbirth is passed. Hopefully this will result in the delivery of a live and vigorous calf that stands and suckles without delay. This may take some time, particularly in heifers that have not calved before, and patience may be required – although it is

The successful culmination of a year's care and attention.

important to recognise when labour is not progressing normally so that the cause can be investigated and timely assistance provided if necessary.

NORMAL CALVINGS

During first-stage labour, which may continue for several hours, the calving cow will often appear a little agitated and move away from the herd. It is important not to intervene too early during first-stage labour; even moving the cow from a familiar calving paddock into an unfamiliar calving box will frequently delay the process, impair cervical dilation, and result in a more difficult delivery than necessary.

When I was a young and inexperienced vet I remember asking an older colleague why, when called to a difficult calving out of hours, I ended up struggling far more than he did. 'Well, my boy,' came the answer, 'it's because you are always in too much of a rush; the 'phone goes, you jump up and drive like a loony. You need to take more time!' How could this make any sense when faced with an emergency situation and a difficult calving? 'When I get a call at 2 o'clock in the morning...' the wise old vet continued, 'I acknowledge the urgency of the situation to the farmer and reassure him I am on my way; then I sit on the edge of the bed and smoke a cigar. After getting dressed I boil some water and make a cup of coffee. I then drink the cup of coffee and smoke another cigar. Only then do I drive slowly and carefully to the farm to provide the requested assistance.'

Some of the calvings my colleague was called to were difficult, of course, but many received a comment from the farmer about what a good vet he was because he made the calving look easy. He explained this to me: 'That...' he said, '...is the importance of not rushing to a difficult calving, to allow

the cervix time to fully dilate!' (by which he meant that only then could second-stage labour proceed – although when called to a calving where the calf is stuck halfway out, speed of attendance is important for both calf survival and the ability of the dam to stand after the delivery).

Second-stage labour involves regular and powerful uterine contractions, accompanied by conscious contractions of the muscles of the abdominal wall, which rupture the foetal membranes (the 'water-bag') and push the calf out. This should only take an hour or two, possibly less, for a cow, and up to about four hours for a heifer. Early during second-stage labour the water-bag, which may be up to the size of a football, will appear through the vulval lips. When this bursts, both front feet should appear, followed by the calf's nose lying on top of the forelegs (assuming the calf is in a normal presentation). The

A cow just entering second-stage labour.

calf should be born soon after the nose is first seen.

If second-stage labour does not seem to be progressing as expected, or the calf's legs and nose do not appear at the vulva, the cow should be quietly restrained and the situation investigated. A crush might be a suitable means of restraining the cow to allow this investigation, but is not a suitable place to continue with the delivery of the calf. The investigation will involve inserting a cleaned, gloved and lubricated hand and arm through the vulval lips and into the vagina to confirm that the cervix is fully dilated and that the calf is in a normal presentation. If all is well you might choose to leave well alone, but assistance can be given: cleaned and disinfected, soft calving ropes should be looped, one round each of the calf's forelegs above the fetlocks, and then, using copious amounts of obstetrical lubricant and working with the cow using either manual traction or a calving aid (to prevent losing progress that has been made, rather than to pull the calf out), the calf may be gently and patiently delivered.

Following delivery of the calf, delivery of the foetal membranes, the 'afterbirth', would usually be expected within a fairly short period of time to complete third-stage labour, certainly within four to six hours. Failure to deliver the afterbirth within such a period is often seen in cows that have delivered twin (or even triplet – it happens but is exceedingly rare!) calves, or which have suffered milk fever or some other complication during calving. This will delay the start of uterine involution and repair, and will often result in a post-partum infection, both of which will, in turn, have an adverse effect on the cow's

Assistance being given to a calving cow using good old-fashioned man-power!

A calving aid in use. The author prefers to use T-bar calving aids because, although more likely to slip than 'Vink' models (though perhaps only if they have been poorly positioned or if too much tension is being applied), they will fall off if the cow starts jumping around, rather than smacking into the operator's legs!

return to ovarian cyclicity and the time it takes her to conceive again.

The administration of oxytocin between six and twelve hours after calving to cows that fail to 'cleanse' may be sufficient to stimulate further myometrial contraction to expel the cleansing. If not, assuming the cow remains well in herself, there is no urgency to remove the afterbirth, which may well drop out unaided a few days later. If it does not, or if the cow becomes sick, veterinary attention should be sought to remove it and to treat any infection that may be present. (Even if it is passed spontaneously, an argument can still be made that any cow that has 'held

her cleansing' should be examined by a vet and treated appropriately to maximise her chances of conceiving as early as possible during the subsequent bulling period.)

ASSISTED CALVINGS

When second-stage labour does not progress as expected, one of a multitude of situations might be the cause: there may have been a failure of the cervix to dilate, or a torsion of the uterus may be occluding the birth canal, or the calf may be in an abnormal presentation, or simply too big. The presence of a dead calf may add further complications. The sooner the cause of the problem can be ascertained and action taken to correct it, or help requested, the greater the probability of a positive outcome.

When investigation reveals a live calf in a normal presentation but which the cow is unable to deliver without assistance because of foeto-maternal disproportion – either because the calf is too big or because the cow is too small (or too fat) – a decision must be made without delay about whether to deliver the calf *per vaginum*, or to call for expert veterinary assistance to help with this, or to resort to Caesarean surgery to deliver the calf. Unfortunately there is no exact science that will inform whether to deliver the calf in these situations by traction, or to resort to surgery, although bone impacting on bone (the calf's head impacting on the maternal pelvis), an inability to engage both the calf's legs and its head into the cervix together, and a lack of any progress despite manipulation and traction, would all tend to suggest that surgery may be necessary. Ultimately, however, the decision will largely be made based on experience (if there is any doubt at all the wise course of action is to seek expert assistance), and only with the benefit of hindsight will there be an indication as to whether the correct course of action was chosen.

When traction is chosen, hygiene, the use of profuse quantities of obstetrical lubricant, and sympathetic use of a calving aid will all help to achieve success, remembering that, within reason, it is not how tight the calving is, but how long the calf is stuck in the birth canal that causes most damage. After a traumatic and difficult calving it is obvious that the calf may need extra attention: its head, and particularly its tongue, may be oedematous and swollen because of compromised venous return due to the pressure experienced during its passage through the birth canal. This may affect its ability to suckle the colostrum that it so urgently needs after being born. Intake of colostrum is essential to provide the calf with passive immunity from its dam, and after a traumatic delivery the calf may be acidotic,

causing it to be 'sleepy' and slow to stand (further compromising colostral intake).

In such cases a dose of a non-steroidal anti-inflammatory drug (NSAID), or even a solution of bicarbonate if the calf has been delivered by a vet, might be beneficial. It is also important that while caring for the new-born calf, the dam is not forgotten. Damage to the vagina, cervix and uterine wall should all be ruled out by gently examining the tract internally, water should be offered, a dose of a NSAID given, and the need for calcium considered.

The alternative is Caesarean surgery. This is not necessarily a straightforward option. It is major surgery that is often carried out on a stressed and fractious patient in less than ideal surroundings: a farm building is hardly a sterile operating theatre, and tractor headlights are a poor substitute for operating lights! The outcome, however, can be remarkably good, especially when

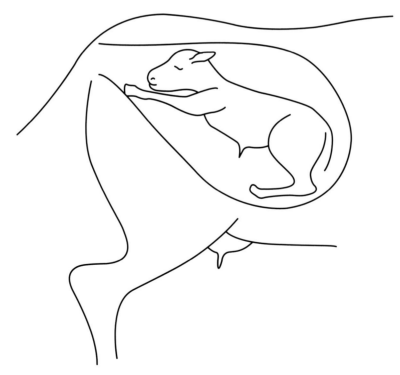

A calf at parturition in a normal presentation.

the decision to progress to surgery is made early, and with a skilled surgeon supported by experienced assistants.

MALPRESENTATIONS

On other occasions an investigation might reveal a malpresentation of the calf to explain the lack of progress. Common malpresentations include calves in a normal 'head first' position but with one or both front legs back; or the head back; or a 'backwards' calf, either with the hind legs presented or with both back legs extending forwards towards the cow's head and only the tail presented – a true breech presentation.

'Head First' with a Single Leg Back

Correcting the presentation of a 'head first' calf with a single leg back should be relatively straightforward, particularly if the leg is

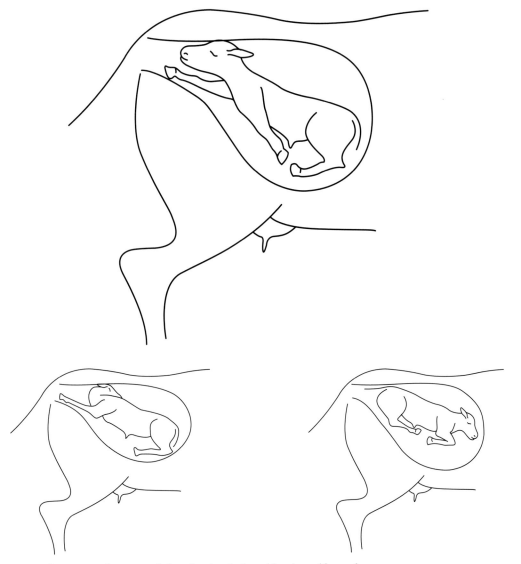

Common malpresentations at calving: leg back, head back and breech.

flexed at the knee so that only the lower part of the limb is pointing in the wrong direction. However, if the entire forelimb is positioned along the calf's chest with the foot towards the cow's head, the knee needs to be brought towards the birth canal by passing a calving rope round the top of the leg and then manipulating it down the limb whilst applying gentle traction to the rope, at the same time as pushing the shoulder back so that the knee pulls up with the upper leg rotating like a fulcrum. Once the knee is accessible it can then be pushed back as the foot is pulled forwards, with a hand cupping the foot to prevent any damage to the uterine wall. (If the foot cannot be reached with a hand, a calving rope looped round the knee can be pushed down the lower limb until it can be placed in the interdigital cleft from front to back. Traction will then flex the foot and pull it towards the birth canal, and the lower limb will straighten as the knee is pushed back.)

Both Legs Back
Where both legs are back but both are flexed at the knee they can be repositioned as above, but where both legs are pointing backwards from the shoulder the situation might be complicated if the calf's head has already been pushed into the birth canal or even out of the vulva. In these cases, where the calf is still alive, there is no option but to push the calf's head back into the uterus (after placing a head rope to ensure that the head can be advanced into the birth canal again after the legs have been repositioned) to allow access to reposition the legs. This can take considerable effort, and great care must be taken not to damage the cow during the procedure. Veterinary assistance can be useful in such cases to administer an epidural anaesthetic to stop the cow straining, and drugs to relax the uterus to facilitate the process. (If the calf is already

dead, removing its head to access its legs might be a less traumatic option for the cow than pushing the head back into the uterus.)

'Head Back'
Correcting a 'head back' presentation can be more challenging, especially if the cow has been trying to calve for some time, resulting in the calf's neck being pushed hard into the birth canal; again, in such cases the administration of an epidural anaesthetic and drugs to relax the uterus may be very useful. The corners of the mouth, the eye sockets or the ear canals may provide useful sites for fingers to grip into when trying to reposition the calf's head, remembering that it may need to be rotated under or around the legs rather than simply being pulled into the birth canal; and take care to protect the uterine wall from being damaged by the calf's teeth (which are sharp!) as the head is pulled round.

A head rope should be placed as soon as this becomes possible: place the point at which the rope slides through the loop to make the noose in the calf's mouth (this helps keep it in place while it is positioned, and then directs the head in a more normal position during delivery as traction is applied), then manipulate the noose over the crown of the calf's head and behind the ears before pulling it tight to keep it in position. Hooks that tighten into the eye sockets as traction is applied can also be a useful tool when trying to straighten a malpositioned calf's head. (When used correctly these are nowhere near as brutal as they may sound, and they can be particularly useful if the calf is already dead.)

'Backwards' Presentation
Identifying a calf in a 'backwards' presentation should not be difficult, because if the feet are visible at the vulva they will be 'the wrong way up' (assuming that the calf is

not twisted or upside down). If the feet cannot be seen, a quick feel will identify the points of the hocks and confirm that the fetlocks and the hocks flex in different directions (the fetlocks and the knees, where the forelegs are presented, both flex in the same direction). If both hind limbs are presented and a decision has been made that a *per vaginum* delivery is to be attempted, soft calving ropes, cleaned and disinfected, should be looped, one round each of the hindlimbs above the fetlocks, to provide purchase to assist with the delivery – as with a calf being delivered in a normal presentation.

However, unlike the situation with a calf being delivered in a normal presentation, there is no gentle increase in diameter as first the calf's legs enter the cervix and then the head moves through with the cervix stretching as the nose and then the crown of the head is delivered: the relatively small diameter of the legs increases rapidly as the bulk of the calf's hind end is pushed into the birth canal. At the first possible opportunity it should be ensured that the calf's tail is tucked down between its legs to prevent the tail-head sticking up and increasing the dimensions of the calf.

A dilemma is then reached, as obviously the calf cannot start to breathe independently until its head has been delivered, and yet its oxygen supply will become compromised as soon as the umbilical cord is squeezed as its abdomen moves into the birth canal. From the calf's point of view a rapid delivery is therefore required. The cow, however, will benefit from a slower, more controlled delivery, and time is well spent ensuring the cervix is fully dilated before attempting to deliver the calf – and plenty of obstetrical lubricant will help!

Breech Presentation

Where the calf is in a backwards presentation with one or both hind limbs extending forwards towards the cow's head this is a true breech presentation, rather than simply a backwards presentation; this is similar to the situation where the calf is in a normal presentation but with one or both fore limbs pointing towards the cow's head, where the knee has to be accessed and then the fetlock so that the lower limb can be 'levered' towards the birth canal. In a breech presentation delivery relies on accessing the hocks and then the fetlocks, and then rotating the lower limbs towards the birth canal using a combination of traction of the foot or lower limb and repulsion of the hock. This might be facilitated by manipulating the lower limb horizontally across the width of the cow's abdomen, rather than attempting it with the calf's lower limb in a vertical orientation – and of course great care must be taken to prevent damage to the uterus from the calf's foot as the manoeuvre is completed.

The Presence of Multiple Calves

As hinted at above, malpresentation of the single calf is not the only reason for second-stage labour not to result in a successful delivery. The presence of multiple calves can block the birth canal, preventing the delivery of any of them without assistance. Investigation of such cases may identify multiple feet and legs, and more than one head. It is usually possible, by carefully feeling the presented limbs, to identify which is paired with which, and then to proceed with the delivery of one of the calves as above, while pushing back the second calf to allow the first calf to enter the birth canal. (As a bonus, where multiple calves are present they are usually not too big, and so delivery, once it has been sorted out which leg goes with which, tends to be straightforward.)

Even after the second calf has been delivered the uterus should still be investigated for the possible presence of a third, however unlikely this may seem (and at the same time the vagina and cervix can be carefully palpated to ensure that no damage has occurred during the delivery). The second calving that I was called to assist with as a qualified vet was a triplet birth – so shortly into my career as a vet, and without my stamina fully developed, any difficult calving was a challenge. I managed, however, to 'untangle' the twins and deliver the first calf, before calling my boss for help to deliver the second. I thought he was making fun of me when, having delivered the second, he said that I had better now get on and deliver the third!

Schistosome Calves

If it proves impossible to determine which leg goes with which when faced with multiple legs all trying to enter the birth canal, applying a rope to one and pulling gently may cause one of the other legs to also move towards the vulva, allowing the pair to be identified. If all the legs move, or none of them moves, then the possibility of an abnormally developed calf, a *schistosomus reflexus,* needs to be considered, especially if a head and a tail can also be identified. Schistosomes are calves that develop with a reflex of their spine, which prevents the body cavities from closing, so that all four legs, the head and the tail lie close together within the uterus, and the viscera are unenclosed within the calf's abdomen and thorax. If these calves present during labour with the bend in the spine accessible then it may be possible to cut this internally and deliver the front and back parts of the calf separately – but if the legs, head and tail are presented then Caesarean surgery is almost inevitable.

A schistosome calf: this can cause considerable confusion to anyone trying to assist with the delivery, which will be challenging and may require Caesarean surgery.

Maternal Causes for Failure of Second-Stage Labour

Primary failure of the cervix to dilate is rare, although it may occur secondary to another problem, including malpresentation of the calf or a uterine torsion that prevents the progress of the calf into the birth canal.

The typical presentation in the case of a uterine torsion is often an older cow with at least a degree of milk fever, which was thought to be going to calve 'last night' but has still not delivered her calf in the morning. A careful and gentle internal examination of the vagina where a torsion exists may reveal a spiralling of the birth canal. If the torsion is only partial it may be possible to pass a hand and arm through the twist into the uterus and grasp one of the calf's legs. If so, it may be possible to 'swing' the calf, perhaps assisted by vigorous external ballottement of the cow's abdomen in the appropriate direction and at the appropriate time, to reposition the uterus; it may then be possible to proceed with the delivery of the calf, although frequently in such cases the cervix will have failed to dilate fully so patience is often necessary.

Many years ago I worked with a vet who, whenever he was called to a difficult calving after about four o'clock in the afternoon, always diagnosed a partial uterine torsion.

GOLDEN RULES FOR CALVING

- Don't intervene too soon, but don't leave it too late. Excessive or premature disturbance can delay or even halt progression through first- and second-stage labour, but a lack of timely assistance can result in disaster.
- Make sure help is immediately available, and know when to call for expert assistance (and do it early!).
- Ensure adequate restraint of the cow to ensure safety both for yourself and the animal. A crush, while useful to allow an initial examination, is not a suitable place in which to calve a cow. A halter robustly fastened (but with a quick-release knot) will suffice, but a hinged calving gate in the corner of a pen is ideal.
- Be clean. Keep your calving aid and ropes clean and disinfected at all times. Wear clean waterproofs. Wash your hands and arms, and the rear end of the cow if necessary, before intervening.
- Use plenty (even excessive amounts) of obstetrical lubricant.
- Remember you may need to push before you pull to correct malpresentations.
- Use your calving aid with patience and care.

Calving aids should be used to help the cow and prevent losing progress that has been made. They should not be used to pull the calf out – and you should never resort to using a winch or, even worse, the towing hitch on a vehicle.
- After delivering the calf always examine the cow internally to ensure no more calves are present within the uterus.
- Don't forget to look after the cow as well as the calf. Offer a drink and consider the possible benefits of NSAIDs and the need for any other medication.

A calving box in the corner of a field with a calving gate in use.

He managed to insert his arm and swing the calf, making suitable grunting and gasping noises while he was doing so to demonstrate the effort required, and correct the torsion. He then expressed concern that the cervix had not fully dilated, and to deliver the calf immediately would risk tearing it, so he advised waiting for a couple of hours and then calling for further assistance. Of course by this time another vet would be on call, and would be duly called up if the calf had still not been delivered – which, of course, it would not have been because the cause of the problem was not a torsion in the first place!

Where a full 360-degree torsion is present it will not be possible to pass a hand into the uterus. In such cases it may be possible to correct the torsion by rolling the cow over her back in the appropriate direction, perhaps while pressure is applied to her abdomen; but success is not guaranteed, and an early decision to proceed to Caesarean surgery may result in a better outcome.

Foetal Monsters

In rare cases the failure of second-stage labour is caused by the presence of a foetal monster or a dead and infected calf within the uterus, and in this instance expert veterinary assistance is always indicated. It may then be possible to section the calf within the uterus and deliver it in pieces *per vaginum* (certainly the preferred option if the calf is dead and infected). As a rule of thumb, however, if more than three cuts will be required to deliver the calf, then Caesarean surgery might be more appropriate (albeit with a guarded prognosis if the calf is dead and smelly).

CAESAREAN SURGERY

When all other options have been exhausted, or to maximise the probability of survival of both the cow and, hopefully, the calf, there may be no other choice but to deliver the calf by Caesarean surgery. Although this is major surgery, attention to several factors, not least an early decision to progress to surgery before both cow and attending veterinary surgeon are exhausted, will have a positive influence on outcome, maximising the probability of cow and calf (assuming that the calf is not already dead) survival, and the future breeding potential of the cow (although the surgical damage to the uterus will obviously need time to heal, which will delay time to conception following the surgery). A clean, dry, well-lit area in which to perform the surgery, with good restraint for the cow and sufficient experienced and well-briefed assistance (ideally at least three people: one to look after the cow, one to scrub up and assist during the surgery, and one to look after the calf) will all help towards a positive outcome.

Surgery is best performed with the cow standing, and so sedation may be best avoided, unless she is particularly fractious. If it is necessary, however, it may be better given early to ensure that the cow is recumbent before surgery is started, rather than surgery being started with the cow standing and then becoming recumbent during surgery with the abdomen open. (If sedation is given don't forget that it will also have an effect on the as-yet unborn calf as well as on the cow.)

If there is any doubt about the cow's ability to remain standing for the entire duration of the surgery, a rope tied to the lower right hind leg and passed under the cow's abdomen so it can be pulled hard by an assistant if the cow does go down during the surgery will hopefully ensure that the left side of the abdomen, where the surgical incision will almost always be, remains uppermost and as minimally contaminated as possible. If a decision is made to operate on a recumbent animal the protrusion of the rumen from the surgical incision, at least until the calf has been delivered, should be predicted and an assistant should be briefed about how best to keep it in place before it happens!

Administering the Anaesthetic

Licensing laws in the UK mean that the choice of local anaesthetic agent is limited to procaine, which is usually used as a 2 or 3 per cent solution with adrenalin added to cause local vasoconstriction at the site of injection – this means that the anaesthetic agent persists at the site of injection for longer than it would otherwise.

Injection pattern is largely dictated by the preference of the veterinary surgeon performing the surgery. Some inject a line of local anaesthetic into the flank of the cow along the line they propose to make their incision: a line block. Whilst this may be perfectly satisfactory, local anaesthetics inhibit healing, so injecting the anaesthetic along the line of the incision seems counter-intuitive, and a large volume of anaesthetic, which is toxic, is required to achieve success. Furthermore, although good skin and muscle desensitisation can be achieved with a line block there is little desensitisation of the peritoneum.

An alternative is to move the line of anaesthetic in front and above the predicted site of the incision: an inverted L block. This has the advantage of moving the local anaesthetic to a site distant to the incision, avoiding any adverse consequences on healing; however, it does require the use of a greater volume of the agent, and retains the disadvantage with respect to peritoneal desensitisation.

The most elegant solution is to perform a paravertebral nerve block: it moves the local anaesthetic agent to a site distant to the incision, it minimises the amount of local anaesthetic agent used, and it also provides for some peritoneal desensitisation. In this method the segmental nerves supplying the relevant dermatomes of the left abdominal wall are 'blocked' as they leave the spinal cord by injecting local anaesthetic agent above and below the ligaments caudal to the transverse processes of the last thoracic vertebra (T13) and the first three lumbar vertebrae (L1, L2 and L3). As well as a loss of sensation of the left flank, success can be predicted by a bowing of the vertebral column to the left as muscle tone is lost in the left flank, and by an

The locations required to achieve successful paravertebral anaesthesia, with the local anaesthetic agent required to be injected both above and below the ligament between the lumbar vertebral transverse processes in each site.

The bowing of the spine towards the desensitised left flank of the cow, confirming successful paravertebral anaesthesia.

engorgement of the small blood vessels in the skin as their muscular walls relax.

Preparations Before Surgery

After the local anaesthetic agent has been administered, the final preparations of the cow can be made for surgery. Any calving ropes still attached to the calf and protruding from the vulva should be removed and the perineum cleaned. The cow's tail should be tied to one of her hind legs to prevent it from flicking into the wound during surgery, or an epidural anaesthetic should be administered. Premedication with a NSAID and antibiotic of the surgeon's choice should be given, and a generous surgical site (it is difficult to make it bigger once surgery has commenced!) should be clipped, washed, scrubbed with surgical scrub and rinsed with surgical spirits. From

now on the left flank of the cow should not be touched.

It then comes to the surgeon to prepare him- or herself for the surgery. As well as the necessary surgical kit, a sterile calving rope or two, embryotomy wire and lint-free sterile towels should be to hand. Surgical drapes are a matter for personal preference; they may be regarded as best practice but many veterinary surgeons operating on cattle do not use them. A clean gown, however, has been shown to have a positive influence on case outcome. Then, after scrubbing up, surgery can begin.

Commencing Surgery and Delivering the Calf

The surgical incision is made in the left flank behind the ribs and in front of the hind leg after ensuring desensitisation of the region.

Preparation and planning before surgery is vital.

When the peritoneum has been penetrated, air will be sucked into the abdomen; beware not to advance too quickly and perform an unplanned rumenotomy! Investigation of the cow's abdomen will then hopefully enable one of the calf's hind legs to be identified within the uterus, assuming a normal presentation, which can be drawn to the abdominal incision, allowing the uterus to be exteriorised before being incised to minimise abdominal contamination.

After placing the sterile calving ropes on the calf's legs, the calf can be pulled from the abdomen, after which it becomes the responsibility of one of the assistants; the surgeon's attention should remain directed towards the cow to close the uterus (the placenta and foetal membranes should only be delivered with the calf if it is already free within the uterus, otherwise it should be left in place to be expelled from the vulva later) and the abdomen using suture materials and patterns dictated by personal preference.

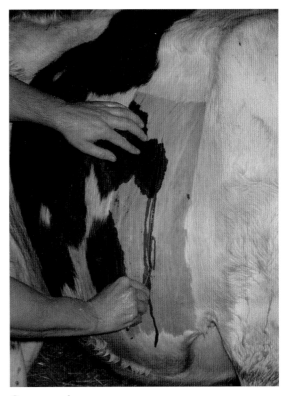

Commencing surgery.

Following surgery, food and water should be made freely and easily available to the dam, oxytocin may be given, and antibiotics and NSAIDS should be continued as directed by the surgeon. Sutures should be removed two or three weeks later.

CARE OF THE NEW-BORN CALF

Birth is a shocking process! The calf, having been protected from extremes of weather and most infectious challenges within the uterus where all its metabolic requirements are provide by the dam via the placenta,

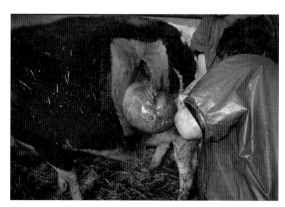

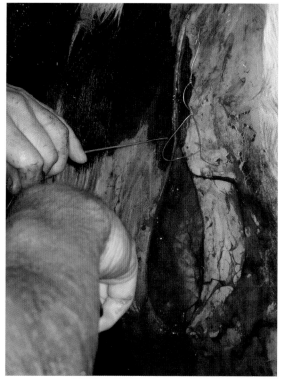

After locating the calf's foot within the uterus and exteriorising it through the abdominal incision (top left) the uterine wall is incised and the calf delivered (top right); the uterus is then closed (bottom left) and then the abdominal wall and the skin (bottom right).

It doesn't matter what the inside of the cow looks like (of course it does!), your surgery will be judged, as well as on whether the calf and cow survive or not, on how neat it looks: your suturing, I was once told, is like signing your name across your work!

is squeezed into a potentially inhospitable environment and exposed to a potentially overwhelming pathogen load without pre-existing immunity (maternal antibodies cannot cross the bovine placenta to ensure protection as the calf is born), and at a time when the lungs need to become functional for the first time and the circulatory system has to adapt to breathing independently.

The calving environment needs to be clean, dry, sheltered and well bedded with fresh straw if it is indoors to protect against inclement weather and minimise infectious challenge.

The calf's breathing needs to be prioritised as soon as it has been born, although patience at the end of second-stage labour to ensure that the umbilical cord is not torn

too quickly can buy some time. Rubbing the calf vigorously with a handful of straw, mimicking the licking of the dam following an unassisted calving, will help stimulate to calf to breathe, and tickling the nasal mucosa by inserting a stalk of straw into the nostrils will add to this stimulation; it might also provoke the calf to cough, which will help expel mucus from the airways. Some farmers will hold the calf upside down by its hind legs or hang it over a gate with its head down to help drain fluid from its lungs.

In reality, any fluid that might be seen draining from the mouth during such procedures is likely to be from the stomach (the volume of foetal fluid in the lungs, which were inactive while the calf was in the

The consequences of a difficult and prolonged delivery: a recumbent and depressed calf with a swollen head and tongue, which may compromise breathing and will certainly delay suckling and colostrum intake.

uterus, will be minimal and will be quickly absorbed into the calf's bloodstream through the alveolar epithelium following delivery); furthermore with the calf hanging upside down, the weight of the abdominal viscera pushing on the diaphragm is likely to impede respiration rather than facilitate it. Pumps and medicines to dilate the airways and stimulate respiratory effort are also available, but these should not be expected to result in miracles: a dead calf is a dead calf, although they may tip the balance in the case of a calf teetering on the brink between life and death towards life.

Immunity, an energy source and hydration are also vital to the wellbeing of the new-born calf; colostrum and the three Qs – quality, quantity, quickly – spring to mind.

Colostrum, the first milk produced by the cow, provides essential nutrients and antibodies to the new-born calf. Unfortunately it is not always possible to assess colostral quality accurately by eye, so measurement of specific gravity using a colostrometer or BRIX refractometer, with a target antibody content of 50g/L, should be carried out. Assuming adequate colostral quality, it is then important for the calf to achieve an adequate intake to ensure a good level of colostral antibody uptake through the 'leaky' gut before the tight cell junctions of the intestinal mucosa close (this begins about six hours after birth and is completed within a day or two, although after this colostral antibodies can still have a valuable protective effect within the lumen of the

The importance of an early and adequate intake of good quality colostrum in ensuring calf health and viability cannot be overstated.

calf's gut). Various authorities have defined 'adequate': six pints within the first six hours of life, followed by a further six pints before twenty-four hours of age, four litres within the first four hours of life followed by another four litres before twenty-four hours of age – or 10 per cent of the calf's bodyweight within five hours of birth, followed by another similar volume before reaching a day of age. This all sounds a lot more complex than it needs to be: it can be summarised as 'a lot, quickly'!

The adequacy of passive colostral antibody transfer can then be assessed by measuring various parameters (IgG content, total protein, zinc sulphate turbidity – targets >10mg/mL, >5.5g/dL and >21 units respectively – or others) in blood samples collected from calves under a week of age.

Irrespective of the adequacy (or otherwise) of passive colostral antibody transfer, if the pathogen challenge is excessive, the passive immunity of the calf will still be overwhelmed. Hygiene in the calving and post-partum environment is important to limit this challenge as much as possible. It is also important to treat the calf's navel, one of the 'weak spots' through which pathogens might gain access. Many farmers will do this with an antibiotic aerosol spray, but an application of a solution of strong iodine BP will dry as well as disinfect the navel, and will reduce antibiotic usage on farm.

CHAPTER 7

CALF HEALTH

Having spent time to ensure bull fertility, getting cows pregnant and then delivering live calves, all that effort will be wasted if the health of the calf is not prioritised to ensure not only its survival but also its performance. The importance of hygiene in the calving environment, an early and adequate intake of good quality colostrum, and disinfection of the navel have all been covered above alongside calving management. Despite best intentions, however, problems can and will still, on occasion, arise.

Navel infection – known as 'navel-ill' – whilst potentially serious in itself and therefore deserving of timely treatment, usually including a course of an appropriate antibiotic, can often act as a source of bacteraemic spread of pathogens to other sites in the body. This will result in peritonitis, liver abscessation, joint infections

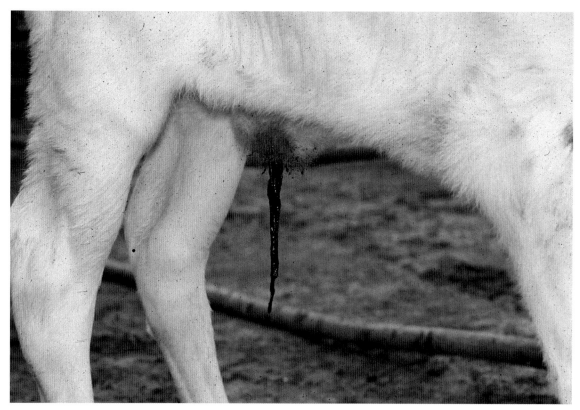

The navel of a new-born calf sprayed with an antibiotic aerosol to prevent infection. Dipping the navel in a solution of strong iodine BP would have helped dry the navel as well as acting as an antiseptic while limiting perhaps unnecessary antibiotic use.

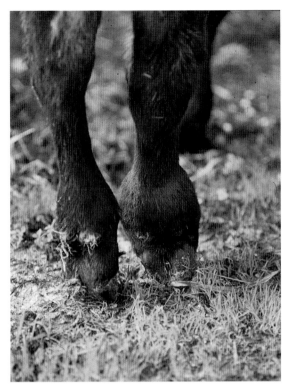

Joint-ill, frequently seen in young calves as a result of bacteraemic spread from an infected navel.

('joint-ill') and spinal abscesses, which may manifest later in the animal's life at a few months of age as a progressive weakness of the hind limbs.

Perhaps the two most significant challenges to the health and wellbeing of young calves are 'scour' (diarrhoea) and pneumonia, particularly in housed calves before and after weaning on a bucket-rearing system, although suckled calves can also be affected.

ENTERIC DISEASE (SCOUR)

Calf scour is a multifactorial condition involving complex interactions between the immune status of the animal, environmental conditions and a plethora of infectious agents, including viruses, bacteria and protozoa. (Parasitic gastro-enteritis is, in this context, considered separately elsewhere within this text.)

The immune status of the young calf is, as mentioned previously, entirely

A scouring calf.

dependent on the passive transfer of colostral antibodies from the dam during the first hours of life. The antibody content of the milk reduces rapidly after parturition and after the calf has suckled for the first time. In addition, the ability of the calf to absorb antibodies from the colostrum also quickly declines after birth as the tight-cell junctions between the cells forming the gut wall become functional. The suckler calf, however, has a significant advantage over the bucket-reared calf in this respect, because although the antibody content of the milk it receives and the ability to absorb those antibodies will both decline, antibodies will still be present in the milk and therefore in the lumen of the gut of the calf, where they will continue to have at least some local protective effect.

The Importance of Hygiene

Environmental contamination can provide a significant threat to the health of young calves and so hygiene is of the utmost importance. Bugs build up during the calving period, so calves born later during the calving period will be exposed to a higher infectious challenge than calves born earlier in the calving period (one of the reasons why a short, compact calving period is so important), so calving sheds, yards and pens need to be kept as clean, dry and well bedded as possible. Even when calving takes place outside, clean, dry paddocks are essential: if the paddock is wet and over-stocked and the cows are covered in mud, the first mouthful a calf will experience after being born will be the mud and the associated bugs coating the outside of the cow's teats rather than the dam's warm, nutritious colostrum full of antibodies. Even if paddocks are not over-stocked and are clean but are wet and the weather is cold, this, even without the bugs, will have an adverse effect on the vigour of each calf, the time it takes to stand and

suckle after being born, and the amount of colostrum it ingests.

The more common infectious agents associated with calf scour include rotavirus and coronaviruses, *E. coli* and salmonella, and cryptosporidia and coccidia.

Rotavirus and Coronavirus

Rotavirus and coronavirus are ubiquitous. Infection is via the faeco-oral route. Rotaviruses primarily attack the lining of the jejunum, affecting the ability of the calf to digest lactose and reducing the absorptive capacity of the gut. This causes, classically, a bright yellow scour, sometimes containing flecks of blood, in animals most frequently between ten days and three weeks old. Coronaviruses, similarly to rotaviruses, attack the lining of the jejunum but cause more severe damage, which extends into the ilium, caecum and proximal colon; this results in an often more profuse scour than is caused by rotavirus, usually affecting animals between two and four weeks of age.

Treatment and Prevention

Differentiating between rotavirus and coronavirus scour is relatively easy and can be done 'cow-side' using one of the lateral-flow ELISA test kits available, but this is largely academic: treatment in both cases is similar and relies predominantly on maintaining hydration and acid-base balance. In the most severe cases this may require fluids to be given intravenously, but this should always be followed by oral fluids, and if the calf remains able to stand, oral fluids (ORT) alone should suffice.

Which ORT used is largely a matter of personal preference: some claim superiority because of an increased energy content (although none provides sufficient energy to maintain growth during the period of illness, so whilst it may have been common in the past to withhold milk from scouring calves, this is no longer advised); some because of an increased bicarbonate or bicarbonate precursor content to help address the acidosis that will be present as an almost inevitable consequence of the scour; some because they contain glutamine to assist with the regeneration of the lining of the gut; and others because of a component that may adsorb toxins. The truth is, the one that is used most, works best: the fluid that is being lost from the rear end has to be replaced, plus 10 per cent, at the front end, which may require three or four 2ltr ORT feeds a day in addition to any milk that is suckled. If this is achieved, renal perfusion will be maintained and the kidneys will sort out both fluid and electrolyte balance: 'the dumbest kidney'

A young calf being given ORT by stomach tube to correct fluid and acid:base balance.

as they say, 'is cleverer than the cleverest physician'!

Antibiotics will have no effect on viral diarrhoea, although they are often included in treatment plans to 'prevent secondary bacterial infections'. Justification could be made for the administration of NSAIDs to scouring calves to reduce the inflammation in the gut and to make the affected animals feel better and therefore keener to suckle, which will provide fluid, nutrition, and at least some antibodies to act within the lumen of the gut.

Preventing viral scours relies on hygiene and good management of colostral antibody transfer (colostral antibody levels can be boosted by using one of the excellent vaccines available for this purpose to vaccinate the cows a month or two before calving).

E. Coli
Although myriad strains of *E. coli* exist within the environment, it is usually only a few specific strains of enterotoxigenic (ETEC) and enteropathogenic (EPEC) *E. coli* - often with the K99 or other adhesin antigens (which are detected by many of the rapid diagnostic tests available) that facilitate adhesion between the bacterial cells and the cells lining the gut wall - that cause problems. Diarrhoea caused particularly by ETEC usually affects calves within the first few days of life. Onset is usually rapid and the diarrhoea usually severe, and losses are frequently high.

Treatment and Prevention
Affected calves should be isolated for treatment; fluids are vital, and in these cases the use of an appropriate antibiotic is indicated. Prevention requires attention to hygiene, vaccination of the cows to increase colostral antibody levels, and colostral transfer management.

A collapsed calf receiving i/v fluid therapy.

Salmonella

A range of salmonella strains have been isolated from cattle, with S. Typhimurium and S. Dublin predominating. S. Typhimurium is perhaps more associated with causing diarrhoea and dysentery in cattle of any age, sometimes with significant mortality. S. Dublin more often grumbles along 'under the radar' in endemically infected herds causing the occasional abortion, a generalised debility, some pneumonic signs, and perhaps soft, sloppy faeces rather than a classic scour.

Treatment and Prevention

Control depends on hygiene, and sometimes vaccination. Antibiotics may improve the situation but can be a double-edged sword; their use may prolong pathogen excretion, and bacterial resistance is widespread.

Cryptosporidia

Cryptosporidia are single-celled protozoal parasites that reproduce within the cells lining particularly the ilium but also the large intestine. These cells line the part of the gut with the greatest absorptive capacity for fluid, and it is the damage caused to them as the new generation of parasites bursts out of them, that accounts for the profuse, greenish diarrhoea produced by affected animals. Infection is via the faeco-oral route, and clinical disease can be seen at any age from one week onwards. Morbidity is often high, and although infections are often self-limiting, losses from cryptosporidia can and do occur.

Treatment and Prevention

Sulphur-based antimicrobials may have some treatment benefits, but newer antibiotics, Paromomycin for example, are more efficacious – although they are subject to the drive to reduce antibiotic use in general. Dosing with the specific antiprotozoal agent, Halofuginone, may therefore be preferred as both a treatment and preventative measure, but in this case the therapeutic dose is close to the toxic dose, so care needs to be taken to dose calves accurately; furthermore it needs to be understood that the treatment itself may make calves feel unwell before they improve. Prevention, as is so often the case, revolves around good hygiene, particularly in calving sheds, yards, pens and paddocks.

Coccidia

Coccidia, another protozoal parasite, are frequently regarded as a pathogen of housed, weaned, bucket-reared calves, but recently cases of poor performance associated with sloppy faeces, or diarrhoea affecting older suckled calves at grass caused by *Eimeria zuernii*, have been described. Care must be taken to differentiate this from the problems that may be caused in slightly older calves, including diarrhoea, by gastro-intestinal nematode parasites, rumen fluke, and the often sporadic, usually fatal condition of unknown aetiology (the involvement of clostridial bacteria has been postulated but not proven), necrotic enteritis.

It must be remembered, when treating scouring calves, that some of the possible causes, particularly some *E. coli* species, Salmonellae and *Cryptosporidiun parvum*, are zoonotic and can affect humans. High standards of personal hygiene are required at all times when in contact, both direct and indirect, with farm livestock, and particularly when disease is present amongst those livestock.

PNEUMONIA

As with scour, pneumonia is also a multifactorial condition involving complex interactions between the immune status of the animal, the environment and infectious agents. Although perhaps most commonly associated with housed, weaned, bucket-reared calves, suckler calves that are outside on their dams can also be affected, and pneumonia can be a significant problem in older suckled calves after weaning and housing.

Colostrum remains important in the prevention of pneumonia despite the frequently prolonged period between birth and any problem being seen. Published data shows that calves with a poor colostral antibody status, and those that scour, are more likely to also suffer from pneumonia than herd contemporaries that achieved a good colostral antibody status and did not scour. Abattoir surveys also show that pneumonia is more common amongst fattening cattle than is often recognised: one third of the fat cattle slaughtered for human consumption have areas of consolidation within their lungs typical of pneumonic damage, and this will have compromised both welfare and performance, despite an absence of any recorded illness or treatment.

The Risk Factors for Pneumonia
Managing the Housing Environment
Whilst most spring-born suckler calves are kept outside, making any control over the environment almost impossible, the weather can still have an influence on the prevalence of disease. In particular, temperature fluctuations between warm days and cold nights seems to be a risk factor for particularly pasteurella pneumonia. The major risk period, however, for pneumonia in spring-born suckler calves, is in the autumn after weaning and housing. Adequate ventilation is, of course, necessary to remove dust and noxious gases (and if the respiratory tract defence mechanisms are overwhelmed removing dust from the lungs, they will be less able to remove pathogens efficiently),

moisture (pathogens survive better in moist air) and pathogens. Stocking densities can be critical, particularly where buildings rely on the stack effect to ventilate naturally.

The stack effect relies on the heat that is produced by the animals in the building warming the air, which rises and exits the building, often through a ridge vent, sucking in fresh air through inlets, usually above the walls at the sides of the building. For this to work satisfactorily the building needs to be adequately stocked (too low a stocking density can cause as many problems as too high a stocking density!), and there needs

A poorly ventilated (and poorly lit) calf shed with vegetation growing up the outside of the building, reducing the air inlet area.

Modifications made to improve the air inlet area and ventilation to try to ensure calf health.

CALF HEALTH **105**

to be a sufficient air outlet and inlet area – how often do you see the air inlet area compromised by the addition of a lean-to on the side of the rearing shed, by bales of straw or silage being stacked against the side of the building, or by trees, brambles or ivy growing close to or up the side of the building? There also needs to be a sufficient height difference between the two, and a sufficient temperature gradient between the inside of the building and the ambient outside temperature.

To reduce the humidity within cattle sheds further they should be well drained and cleaned out regularly, and water troughs should be carefully sited. Regular cleaning out will also prevent the build-up of bedding, which will reduce the air space per animal within the building, effectively increasing the stocking density of the building.

The Infectious Processes Involved in Calf Pneumonia

The conventional view of the infectious processes involved in calf pneumonia is of a primary viral challenge, possibly from Parainfluenza 3 (PI3) disrupting the cilia in the major airways, compromising the action of the muco-ciliary escalator and its efficiency at removing infectious agents and other debris from the lungs; or from respiratory syncytial virus (RSV) 'blowing holes' in the lung tissue resulting in bulla formation, followed by secondary bacterial invasion. However, many of the bacterial pathogens with a known involvement in the calf respiratory disease complex – *Mannheimia haemolytica*, *Pasteurella multocida* and *Histophilus somni*, for example – can, and do, cause pneumonia as primary pathogens (although their involvement in mixed infections is common). Mycoplasmas, a half-way house between viruses and bacteria, particularly *Mycoplasma bovis* and

Mycoplasma dispar, are also not uncommonly involved.

Treatment

Treatment of such a complex condition is challenging. Where there is bacterial involvement, either as the primary pathogen or as a secondary invader, the use of an antibiotic is probably indicated – but which one, and which animals should be treated? A wide range of antibiotics are licensed for the treatment of calf pneumonia, and in the correct circumstances can be very effective, but many of these have now been designated by the World Health Organisation as 'high priority, critically important' (HP-CIAs) in human medicine, and should only be used to treat farm animals as a last resort when all other treatment options have failed (by which time the damage to the lungs of the affected animals might be so severe that even HP-CIAs will fail).

Which Antibiotic?

Consideration needs to be given to the spectrum of activity of the antibiotic chosen (many strains of Mannheimia and Pasteurella are now known to be resistant to Oxytetracycline, one of the antibiotics commonly used to treat pneumonic calves, and penicillin and its derivatives will have no action against the mycoplasmas because of their lack of a cell wall) and the ability of the medicine to penetrate diseased tissue: if it can't get to where it is needed it will not work (remembering that in many chronic cases the damage to the lung tissue will already be beyond effective treatment).

Which Animals to Treat?

Whether to treat only the affected animal(s) within a group can also present challenges. Prophylactic treatment is now frowned upon within 'responsible use' guidelines, but not all affected animals are obvious even to the most experienced stockman, and metaphylactic use may have not only health and welfare benefits, but may also reduce the overall antibiotic use during a disease outbreak. Guidance on which animals to treat can be gained by taking the temperature of all the animals in the group at regular intervals, but this is time consuming and stressful to the animals (and so might exacerbate the problem). A useful alternative might be to change from individual treatment to group treatment based on the proportion of animals in the group showing clinical signs (if more than one third of the animals in the group are affected, group treatment might be indicated), or on food consumption: if this drops by 10 per cent or more, again, group treatment might be indicated.

It is also important to recognise that antibiotics will have no effect on viral pathogens, yet calves affected by viral pneumonia still need treating to limit the damage caused within the lungs, and to address the pain caused by pneumonia. In this respect corticosteroids have been widely used by some in the past, but their use has now been largely replaced by the use of non-steroidal anti-inflammatory drugs (NSAIDs); their use as part of the treatment for pneumonic calves should be considered essential rather than optional.

Vaccination
Prevention, of course, is always better than cure. Although it will never negate the importance of environmental management, vaccination to ensure the immune status of the calf is a commonly used tool in the prevention of calf pneumonia. Unfortunately there is no single vaccine, nor any combination of vaccines that will protect against every pneumonia-causing pathogen, and even if a relevant vaccine is chosen, if the infectious challenge is overwhelming, problems are still likely to occur.

As regards which vaccines to use and in what combination, consideration needs to be given to the pathogens included, the route of administration, the time-lag until onset of protection, the duration of cover and the cost-benefit, all in the light of the age and status of the animals requiring protection, and the age from which protection is required. In general, intra-nasal vaccines, of which several are available to protect against PI3 and RSV, will have a more rapid onset of action but provide a shorter duration of immunity than many of the injectable options; the latter often require two doses of vaccine to be given, with an interval of three or four weeks between doses, before full immunity is achieved – although single-dose injectable Pasteurella/Mannheimia vaccines do exist. Furthermore the injectable options may provide protection against a wider range of pathogens.

In many texts bovine herpes virus 1, the cause of infectious bovine rhinotracheitis, and bovine virus diarrhoea virus, which causes immunosuppression, are included in discussions of the calf pneumonia complex. In this book these pathogens, along with lungworm, will be discussed elsewhere as specific entities.

CHAPTER 8

INFECTIOUS DISEASES

In contrast to the situation in the dairy herd where multi-factorial diseases are arguably the major threat to animal health, welfare and productivity – mastitis and lameness, for example, involving complex interactions between the cows, their environment and a range of infectious agents – in the suckler herd many of the single agent infectious diseases assume a greater significance. Those of arguably greatest significance in the breeding herd will be discussed here grouped by pathogen type: viruses, bacteria, and so on.

VIRAL DISEASES

Bovine Viral Diarrhoea Virus (BVD)

The BVD virus is the cattle-associated pestivirus, a large group of small, enveloped RNA viruses that have the unique ability, when infection of a susceptible, pregnant dam occurs during the first trimester of gestation, to cross the placenta and infect the foetus, resulting in immune tolerance and the birth of a persistently infected (PI) calf (assuming that the pregnancy is not aborted).

BVD has a worldwide distribution (with the exception of the Antarctic continent because there are no cows there!): it is found wherever cattle can be found in the world. Four types of the virus are recognised: type I, type II and type III, which has recently been described

in Italy, and HoBi-like viruses that have been described in North America, and it is closely related to Border disease virus, the sheep-associated pestivirus. Each type of the virus exists as multiple subtypes, each with a differing pathogenicity, and each subtype exists in two biotypes: non-cytopathic, the usual biotype encountered, and cytopathic, which occurs as a spontaneous mutation of the non-cytopathic biotype in PI animals, resulting in the invariably fatal mucosal disease.

Acute infection with BVD often produces little in the way of clinical signs – perhaps a slight pyrexia, but often nothing obvious to see. The most significant consequence, ignoring the effect of the virus on reproductive performance and on the foetus, is the effect on immune function: white cell numbers can be reduced by anything up to between 50 and 90 per cent, depending on the type and subtype of the virus involved. White cell function is also compromised. Together the result can be a profound immunosuppression allowing other infectious agents that may be present to cause more, and more severe disease than they otherwise might; the consequence of an active BVD infection in outbreaks of calf pneumonia is well known, but it doesn't matter which other infectious agent is chosen – for example *Fusobacterium necrophorum*, the cause of foul in the foot. The consequence of infection concurrent with an infection with the BVD virus will be more severe disease than in the absence of BVD. (Active BVD infection has even been linked with the spread of *M. bovis*, the cause of bovine tuberculosis.)

On occasion, however, and particularly (although not inevitably) where type II strains of the virus are involved (the most commonly encountered strains of the virus within GB are type I strains; type II strains are more prevalent within mainland Europe and on the North American continent), more acute disease, involving bleeding due to a viral-induced thrombocytopaenia, may be seen and mortality rates may be significant.

Of perhaps even greater significance in the suckler herd is the effect of BVD on fertility and reproductive performance. In the male it can have a deleterious effect on sperm production and function. Shedding of the virus in the semen of fully immunocompetent, non-PI bulls can continue for up to ten or twelve weeks following an acute infection, and some bulls, following an acute infection with the virus around puberty, have even been found to harbour a chronic, life-long infection within their testicular tissue: this is because the specific circulating antibodies raised in response to the infection are unable to cross the blood-testis barrier to eliminate the infection, with the result that every ejaculate of semen such bulls produce contains the virus!

In the female, ovarian function can be affected and luteal progesterone production reduced. When this coincides with conception and early pregnancy there may be reduced viability of the embryo and increased early embryonic loss. Later in pregnancy foetal death may result in mummification, or, later still, in abortion at any stage.

A calf born with cerebellar hypoplasia, demonstrating a typical wide-based stance and a lack of spatial awareness.

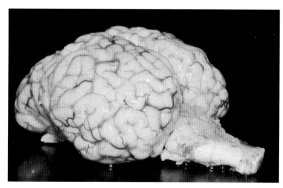

The brain removed from the calf pictured in the image above showing the almost complete absence of the cerebellum.

Foetal loss and abortion are not, however, inevitable. Where a susceptible cow during mid-pregnancy is infected with the virus, with her calf *in utero* consequently also being infected, a variety of congenital abnormalities may result; these particularly affect the central nervous system, which is developing within the foetus at this time. The classic abnormality reported is cerebellar hypoplasia, when the cerebellum, the part of the brain that controls balance and locomotion, fails to develop normally. The extent to which this occurs depends on the exact stage of cerebellar development at the time of infection. The result will vary from a calf that is unable to stand, through to one that appears almost entirely normal. Classically, however, affected animals, if they are able to stand, display a wide-based stance, incoordination, and a star-gazing posture.

The Persistently Infected (PI) Calf

The real majesty of the BVD virus, however, occurs when a susceptible cow is infected during the first trimester of pregnancy before the foetal immune system has developed. The inevitable infection of the foetus means that the developing foetal immune system, assuming that the foetus survives the infection, does not recognise the virus as 'foreign' and so never mounts an immune response to eliminate it. Such animals (PIs) are born infected and remain infected for their entire lives until their inevitable death from mucosal disease (unless they die from something else first!),

characterised by extensive ulceration and erosion of the mucosa of the gastro-intestinal tract and an acute, often bloody, diarrhoea, usually between six months and two years of age (although some will survive long enough to breed, in which case, if they are female, every calf they deliver will also be PI).

In the absence of any immune response to hold the infection in check, PIs become a virus factory generating huge quantities of the virus all day, every day, to challenge and infect any other susceptible animals they are in contact with; it is estimated that 85 per cent of the BVD challenge within cattle herds comes from PIs, with only 15 per cent coming from acutely infected animals, which excrete much lower quantities of the virus for a limited time.

Treatment and Prevention

The measures that need to be taken to protect our livestock from this virus and eliminate infection are therefore obvious: first, to identify and cull all PIs without

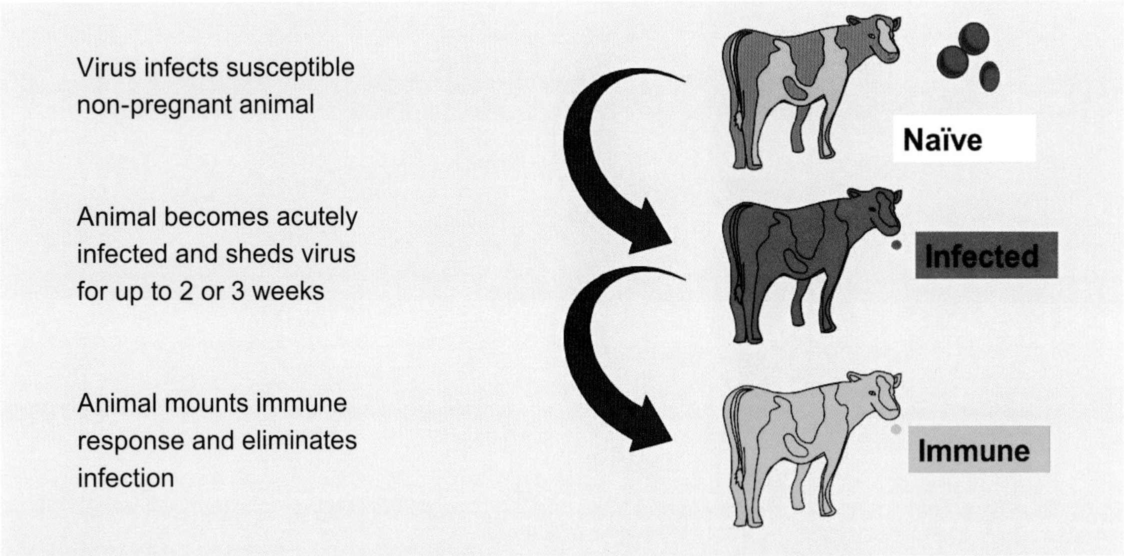

Virus infects susceptible non-pregnant animal

Animal becomes acutely infected and sheds virus for up to 2 or 3 weeks

Animal mounts immune response and eliminates infection

Naïve

Infected

Immune

BVD acting as many other viruses and being eliminated by the immune response of the host in a non-pregnant cow.

Virus infects susceptible cow between conception and 120 days of pregnancy

Naïve

Cow becomes acutely infected and sheds virus for up to 2 or 3 weeks Virus crosses placenta and infects foetus

Infected

Cow mounts immune response and eliminates infection Antibodies cannot cross placenta: Calf remains infected

Immune

PI calf born

When BVD infects a pregnant cow during the first trimester of her pregnancy it will cross the placenta and infect the unborn foetus. Antibodies produced by the maternal immune response to infection cannot cross the placenta, so infection is not eliminated from the foetus: if it is not aborted, it will be born persistently infected and will shed virus every day of its life.

A calf born persistently infected with the BVD virus (foreground) alongside a non-PI herd contemporary born six weeks later.

delay – this goes against advice that may have been given in the past to retain PIs to be mixed with young heifers to act as 'mobile vaccinators' to ensure immunity before they are served: at best it was an inefficient way of ensuring group level immunity, and at worst it caused immune suppression, as described above, ensuring a market for vets for antibiotics and pneumonia vaccines! Second, to ensure all breeding females are immune before conceiving to prevent the birth of further PIs; and third, excellent biosecurity should be ensured to prevent the reintroduction of the virus into the herd.

Excellent laboratory tests, which can be carried out on blood, milk or other tissue – ear notch tissue, for example – are available to detect exposure to the virus and identify PIs. Excellent vaccines are also available to ensure individual animal and herd-level immunity to the virus, although they must be stored and used exactly according to the manufacturer's recommendations, and they must be given to ensure the development of immunity before the cow becomes pregnant; this makes total sense when the aim is to protect the foetus, to maximise reproductive performance and prevent the birth of a PI calf.

The eradication of this virus is entirely within our grasp, so one has to question why we allow it to persist!

Bovine Herpes Viruses

The most commonly encountered of the bovine herpes viruses is bovine herpesvirus 1 (BoHV1), the cause of infectious bovine rhinotracheitis (IBR), an acute and often severe upper respiratory tract infection that, in the suckler herd, often affects recently weaned and housed calves. Other diseases, including infectious pustular vulvovaginitis (IPVV) in the female, infectious pustular balanoposthitis (IPBP) in the male,

both of which can have obvious adverse consequences for conception, and bovine herpes mammillitis are also caused by herpes virus infections and BoHV4 has been linked with post-partum uterine infections. In addition, BoHV1, can cause typical liver pathology in the developing foetus resulting in abortion.

From a business point of view, herpes virus status also has significance for international trade.

The bovine herpes viruses are typical herpes viruses. Following infection, latency within nervous tissue, particularly the trigeminal ganglia in the case of BoHV1, is common or even usual and life-long. Despite usually solid seroconversion eliminating the circulating viraemia, this latent infection allows for recrudescence of the virus at unpredictable times in the future, prompted by uncertain triggers (stress is often mentioned, but what does this actually mean?) resulting in a renewed infectious challenge to other animals in the herd. It should be assumed, therefore, that all seropositive animals retain the potential to present an infectious challenge within the herd. (To complicate the situation, in the absence of recrudescence of the latent infection boosting the antibody response, circulating antibody levels may fall below the current limit of detection, meaning that it is also possible for seronegative animals to be latently infected. It is therefore important to consider herd status as well as individual test results when defining individual animal status.)

Good BoHV1 DIVA (allowing the ability to 'differentiate infected from vaccinated animals'), or 'marker' vaccines, are available for administration via various routes and in both live and inactivated forms to help prevent disease. (In endemically infected dairy herds it has been proposed that vaccination is worth one litre of milk per cow per day, a not insignificant advantage.) However, it must be remembered that vaccination will not necessarily prevent seroconversion to field-strain virus, and the development of latency following challenge and the use of non-marker vaccines (frequently the case in multi-valent vaccines aimed at protecting young calves against multiple respiratory pathogens) and the use of 'live' vaccines resulting in the transmission of vaccinal virus to unvaccinated animals causing seroconversion in these unvaccinated animals, all might compromise future marketing opportunities for the animals involved.

Infectious bovine rhinotracheitis (IBR) can present with a variety of clinical signs, from a mild conjunctivitis with or without a muco-purulent nasal discharge, to a severe necrotic tracheitis, followed by effectively an inhalation pneumonia as the increased respiratory effort caused by the disease sucks the sloughed necrotic tracheal lining deep into the alveoli in the lungs. In these most severe cases death, or slaughter on humane grounds, despite aggressive treatment with antibiotics and anti-inflammatory drugs, is not an uncommon outcome.

The pluck of a weaned suckler calf that has died of acute IBR. Note the severe necrosis affecting the tracheal mucosa, and the pneumonic lung changes as a result of inhaling debris from the trachea due to the increased respiratory effort caused by the disease. The effect of treatment in such cases is frequently disappointing, which is not surprising given the pathology present.

A dairy cow with conjunctivitis and weeping eyes caused by BoHV1.

Malignant Catarrhal Fever

On a herd level malignant catarrhal fever (MCF) is not a significant disease; it is sporadic, usually only affecting individual animals, and is not transmitted from cow to cow. When it does occur, however, it presents as a serious, usually fatal condition with striking clinical signs, justifying its mention here. It is caused by ovine and caprine herpesvirus 2 (and alcelaphine herpesvirus 1, which is a wildebeest-associated herpesvirus), so an association between affected cattle and proximity to

MCF. The clinical features in this case are limited, but the typical corneal opacity is obvious.

sheep or goats, especially during or soon after the lambing or kidding period, can usually be identified.

Infection is often asymptomatic within the natural host, but infected cattle are usually extremely ill, presenting with the following symptoms: a high temperature, depression, anorexia, crusting of the muzzle with a purulent nasal discharge, an exudative dermatitis of the udder and teats, haematuria due to erosion of the uroepithelium lining the urinary bladder, and, pathogenomically, corneal opacity, frequently with an associated swelling and partial closure of the eyelids, tears running down the face,

and photophobia. Treatment of such cases is rarely successful and euthanasia on humane grounds is usually advised.

Foot and Mouth Disease

Foot and mouth disease, caused by various strains of Apthovirus, is a vesicular disease characterised by lesions particularly in the mouth and around the coronary band of many cloven-hoofed species, including cattle, sheep and pigs. It had not been recorded affecting animals within the UK for very many years until a large outbreak in 2001 and a smaller outbreak in 2007. The country is now, once again, regarded free from infection, but it remains notifiable, and any suspicion of the disease must be reported to the competent authority without delay to ensure that appropriate investigations are carried out, and actions taken.

BACTERIAL DISEASES

Mycobacterial Diseases

Two mycobacterial diseases are significant in the beef suckler herd although not because of any direct effect on reproductive performance. These are bovine tuberculosis (bTB), caused by *Mycobacterium bovis* (*M. bovis*), and Johne's disease or paratuberculosis, caused by *Mycobacterium avium* subspecies *paratuberculosis* (Map).

While these diseases differ in their detail, both are typical mycobacterial diseases, with a prolonged and variable incubation period after infection, during which the infected animal may progress through an infectious state to a clinically diseased state, although this may never happen and quiescent or latent infection is thought to be common. The triggers for this progression are poorly understood but include stress (progression to disease, particularly in the case of Johne's disease, is commonly seen after calving), and immunosuppression as a result of a compromised nutritional status, or co-infection with another immunosuppressive agent, BVD or BIV for example.

Both also present a significant diagnostic challenge: of course all diseased animals are infected, but not all infected animals are diseased or even shedding infectious organisms, although they all retain the potential to do this and progress to disease. The diagnostic challenge faced by clinicians is to be able to predict in which non-diseased, infected animals infection will progress, and when they will start to shed the infectious organism, creating a threat of infecting other animals they are in contact with and becoming epidemiologically important.

Bovine Tuberculosis (bTB)

BTB, which is under statutory control in the UK and many other countries around the globe, is essentially a respiratory disease often resulting in the formation of abscesses within the lungs and the lymph nodes associated with the respiratory tract (although lesions as a result of infection can be seen in a range of tissues including liver, kidneys and brain); it also often causes a cough and weight loss (although clinical disease is now rarely seen due to statutory testing protocols).

Infection via inhalation is common, but oral infection can also be significant either via the faeco-oral route in older animals, or from feeding milk from infected cows to young calves. Wildlife, and particularly badgers, are often attracted into the vicinity of cattle by the easy access that may be provided to novel food sources, such as concentrates, maize silage and mineral licks. Wildlife is also acknowledged to be involved in the epidemiology of this disease – but the relative importance of undisclosed, endemic infection within the herd and wildlife vectors of the disease are widely debated and likely to differ in every situation.

Johne's Disease

Johne's disease is caused by *Mycobacterium avium* subspecies *paratuberculosis* (Map), and is a gut disease in which the infectious bacteria slowly multiply within the gut wall, and particularly within the lymphoid tissue associated with the gut wall, following infection. This can occur at any age, although there is an age-associated resistance to infection, so infection most commonly occurs during the first hours, days or weeks of a calf's life. As bacterial multiplication progresses the gut wall increases in thickness, reducing its absorptive capacity. This explains the clinical signs of the disease, which are often not seen until the infected animal is between two and six years of age; as the absorption of nutrients through the gut wall becomes increasingly compromised, the affected animal will lose weight and the concurrent compromise to the absorption of fluid will result in diarrhoea, which may become profuse and often has a bubbly appearance. The disease, however, will be taking its toll

A suckler cow showing the first signs of clinical Johne's disease: weight loss and poor condition despite an apparently adequate diet.

A suckler cow demonstrating the clinical signs of end-stage Johne's disease: profuse diarrhoea, poor body condition, poor coat condition, and dependent oedema (bottle jaw) due to the protein-losing enteropathy the disease causes.

A Hereford x suckler cow in good condition and apparently in the prime of health with her superb British Blue x calf at foot – but how can you be sure of her Map status and any risk she might pose to the future health and performance of the calf?

before weight loss and diarrhoea become apparent; the compromised nutritional status of an infected animal will have an adverse effect on its fertility and milk production, so many animals infected with Map will be culled because they fail to become pregnant at the required time, or because of poor calf performance before they are recognised as having Johne's disease.

Mycobacterial Disease Detection and Control

The detection of a cell-mediated immune response, a delayed T-cell hypersensitivity reaction, is used in the single intradermal cervical tuberculin (SICT) and single intradermal comparative cervical tuberculin (SICCT) or 'skin' tests for statutory bTB surveillance in many countries. The SICCT test is slightly less sensitive but more specific than the SICT test, where sensitivity is a measure of the number of false negative results a test gives – how good it is at identifying infected animals – and specificity is a measure of the number of false positive results it gives – how good it is at confirming the absence of infection in uninfected animals. The SICCT is the predominant test used in the UK, with the SICT test being used in many other countries where a higher test sensitivity is required because of a lower prevalence of infection; antibody production (humoral immunity) was thought to be a late feature following *M. bovis* infection (this is now known not necessarily to be the case).

The relatively poor sensitivity of the SICCT test, which is estimated to be, at best, 80 per cent, means that in herds endemically infected with *M. bovis* one in five infected animals will remain undetected following a single herd test, and, in fact, many authorities estimate the sensitivity of the skin test to be significantly lower than this, with figures as low as only 50 per cent, or even less, being not infrequently quoted. Furthermore the binary yes/no interpretation of the test result at an uncertain point during the long period of progression from infection to disease compromises our ability to detect *M. bovis*-infected animals, and therefore to control and hopefully eradicate bTB.

Antibody tests, carried out on either serum or milk, are used in the detection of Map-infected animals. These are highly specific, so an animal testing antibody positive is highly likely to be infected even if testing for the presence of Map itself by culture, PCR (polymerase chain reaction) or phage testing gives a negative result. Note that it has now been largely disproved that an increased number of 'false' positive test results is due to a 'cross-reaction' with the PPD (purified protein derivative) used in the skin test for bTB. Instead, the more likely explanation for the increased number of seropositive animals detected when samples collected during the anamnestic period following a skin test are tested, is that the PPD used in the skin test has boosted an otherwise undetectable antibody response to a Map infection, making it detectable; this means that more Map-infected animals are being correctly identified earlier than would otherwise have been the case.

Although infected animals will be defined on a yes/no basis depending on the magnitude of the antibody result, this is often interpreted, especially where serial test results are available, to define the risk of an infected animal shedding the infectious organism and progressing towards clinical disease, informing on its future management within the herd, and allowing proactive culling decisions to be made.

Mycobacterial disease detection and control is necessary to direct both biosecurity, to prevent the introduction of (more) infection into a herd, and biocontainment, where infection is already present within a herd, to reduce, as far as is possible, its further spread. Going forwards it is likely a combination of both skin testing will be used and antibody detection to make use of the boosting effect of the PPD used in the skin test to maximise the sensitivity of the antibody test: this will allow the detection of animals infected with either *M. bovis* or Map (or both!) as early after infection as possible so that they can be proactively managed with the aim of reducing the prevalence of infection within the herd.

Leptospirosis

Bovine leptospirosis is caused by various strains of spiral-shaped bacteria of the genus *Leptospira*. Transmission is most common during the summer months whilst animals are grazing, and infection is most commonly seen in herds containing purchased animals and using natural service. Co-grazing with sheep and access to flowing water courses are also considered risk factors.

Infected animals often exhibit decreased milk production – known as 'flabby bag' – which in the suckler herd will have an obvious consequence on calf performance. Reproductive performance will be adversely affected, and infection can result in abortion, frequently between five and seven months of gestation. Calves can be born congenitally infected and be 'difficult to get going'.

The infectious organism, unsurprisingly for one that targets the urogenital tract, is shed in the urine and foetal fluids at birth or following an abortion from both acutely affected and carrier animals. It has zoonotic potential and so presents a particular risk to farmers and stockmen working with infected herds, particularly during the calving period. (When people are infected the usual

The typical consequence of leptospirosis: a late abortion, probably occurring sometime after the infection; confirming the cause may therefore be challenging.

consequence is a 'flu'-like illness persisting for some days, but in very rare cases it may present as a meningitis or even encephalitis, which can even be fatal.)

Diagnosis and Treatment
On an individual basis, the diagnosis of leptospirosis as the cause of abortion can be challenging because abortion frequently lags behind infection by a sufficient period of time to make paired serology following the abortion, in an attempt to demonstrate a rising specific antibody titre, unrewarding. Even submission of the aborted foetus to a diagnostic laboratory for investigation carries a disappointing diagnostic success rate. This does not mean that attempts should not be made to confirm the cause of bovine abortions – and indeed within the UK it remains a legal requirement that all bovine abortions are reported to the competent authority, the APHA, in case it is deemed necessary that the affected animal(s) are sampled and tested to rule out brucellosis as the cause. However, from the perspective of leptospirosis it may be more rewarding to establish the status of the herd, rather than of the individual, by sampling and testing a statistically significant number of randomly selected animals from the herd.

If leptospirosis is confirmed, treatment with a variety of antibiotics can be used to address the immediate situation, and vaccines are available to maintain herd immunity and reduce shedding of the infectious agent by carrier animals.

Campylobacter
The Campylobacter comprise a large group of motile, spiral-shaped, gram-negative bacteria. Perhaps the best known as a cause of food poisoning in humans is *C. jejuni*, frequently linked with poor hygiene during the preparation, and inadequate cooking, of poultry meat. Within suckler herds, however,

or any herd of cows where reproduction is managed by natural service, the pathogen of interest is *C. fetus venerealis*.

C. fetus venerealis is a true sexually transmitted disease: it is spread from bull to cow to bull during natural service, and virgin animals can confidently be predicted to be free of infection. In the cow the organism colonises the anterior vagina, from where it may invade through the cervix into the uterus, possibly resulting in early embryonic loss or abortion if a more advanced pregnancy is present. Elimination of the infection is possible thanks to the body's normal defence mechanisms, and the secretion of specific antibodies in the secretions within the reproductive tract associated with oestrus; also a degree of immunity will be established (although this is neither solid nor long-lasting).

In the bull the organism invades the crypts and crevices high in the prepuce; older bulls with deeper crypts and more convolutions of the preputial mucosa are more likely to be a source of infection persisting within a herd than are younger bulls. Once infected, infection in bulls often persists for life unless they are specifically targeted for treatment with antibiotic-containing preputial washes – and these are not always appreciated by the bull, are hard work, and success is difficult to guarantee. Understanding this makes the importance of maintaining a disease-free herd obvious, and reinforces the need for careful selection and biosecurity when sourcing replacement breeding bulls in particular.

Managing C. fetus venerealis Infection
Where infection has been confirmed within a herd there are different management strategies that can be adopted. One is to stop using natural service and use AI alone to manage the herd's reproduction until natural immunity has eliminated the

bug. This, however, requires high levels of intensive management and is frequently unacceptable to the herd owner. Another is to run young bulls with the herd in an attempt to limit the spread of infection. A third is to vaccinate the herd to maximise immunity. In these situations, however, infection will never be eliminated and herd-level reproductive performance will persist in being disappointing.

In many situations the management strategy adopted is to accept the presence of infection in the 'dirty' herd, close this, keep it away from any other animals on the farm (certainly during the serving period) and accelerate the culling of animals from it while at the same time establishing a 'clean' herd from virgin replacement stock. Over time the size of this clean herd will grow, and the numbers in the dirty herd will decline, until it becomes economically viable to cull the remainder of the dirty herd in its entirety, thereby eliminating infection from the farm.

Salmonellosis

The Salmonellae are a wide-ranging group of gram-negative bacteria that can infect a wide range of species, including man, often causing severe gastro-intestinal disease resulting in dysentery and even death. The two most common cattle-associated Salmonellae are the group B organism, *Salmonella* Typhimurium, and the group D organism *Salmonella* Dublin. The typical presentation of a *Salmonella* Typhimurium outbreak, in common with disease caused by many other strains of Salmonella, is of acutely sick animals, some of which may die despite treatment, with a high temperature and profuse diarrhoea or dysentery. *Salmonella* Dublin, however, frequently has a somewhat different presentation, can be easy to overlook, and yet may have a significant impact on suckler herd performance.

Salmonella Dublin is often found coincident with liver fluke infections, infecting carrier cows in endemically infected herds without causing any obvious clinical signs, although it can be responsible for causing the acute clinical signs that typify the disease caused by other species of salmonella. Even in the absence of such signs, however, it remains a significant cause of abortion within herds of breeding cattle, and of non-specific calf disease and mortality.

Treatment and Control

Although antibiotic treatment is frequently given to infected animals the response is often poor, perhaps because of the widespread existence of resistance to many of the commonly used antibiotics in farm animal medicine, even when it is coupled with other supportive treatment and intensive nursing. It is also possible that antibiotic treatment of animals infected with a Salmonella will prolong the period of shedding and increase the probability of the establishment of a carrier status. The control and prevention of outbreaks of Salmonellosis rely heavily on hygiene, on controlling liver fluke in the case of *Salmonella* Dublin, and on vaccination.

Listeriosis

Listeriosis is caused by the gram-positive bacterium *Listeria monocytogenes*, and is often associated with the feeding of contaminated and mouldy silage. It is a relatively common cause of bovine abortion, but can also cause a focal meningitis and encephalitis. The exact clinical signs of this will depend on the exact site of the infection, but reports include classically walking in circles, and unilateral drooping of an ear, the eyelids and lips, perhaps with protrusion of the tongue and dribbling from the affected side of the mouth, depending on which cranial nerves may or may not be involved in the pathological process.

A Simmental x suckler cow displaying clinical signs typical of listeriosis: unilateral drooping ear, eyelids and lips, and dribbling from the affected side of the mouth.

Bacillus licheniformis is another not infrequently encountered cause of bovine abortion associated with feeding poor quality silage.

Clostridial Disease

With the possible exception of hypomagnesaemia, clostridial disease, in its many guises, remains probably the most common cause of the sporadic loss of grazing livestock within the UK, and possibly globally. All are caused by the toxins produced by the relevant clostridial organism, all of which are rod-shaped, non-capsulated (which is important as one of the features that distinguishes clostridial organisms from *Bacillus anthracis*, the bacterium that causes anthrax, which also usually presents in ruminants as sudden death), spore-forming, gram-positive, highly infectious but non-contagious obligate anaerobes.

The spores form on exposure of the bacteria to oxygen, and can then persist in the environment until conditions promote their development - outbreaks of clostridial disease are often seen after earth is moved, for example after clearing ditches, exposing the spores to the grazing livestock - or in the body, again until a suitable anaerobic environment for their development prevails, perhaps in a bruise following bulling or handling, an injury, or even following injection using dirty equipment and employing inexpert technique or, of course, after death.

Perhaps the most commonly encountered clostridial diseases of cattle are blackleg and malignant oedema, caused by the proliferation of *Clostridium chauveoi* and *Clostridium septicum* respectively, and other clostridial species; although the two diseases are difficult to differentiate without laboratory involvement, mixed infections are common and any difference remains academic since the outcome is the same, frequently sudden death, and any advice given will invariably include the use of a multivalent clostridial vaccine. Aggressive antibiotic therapy with high doses of penicillins may successfully save the animal's life, but it is rare to identify affected animals sufficiently early, due to the frequently per-acute nature of the disease and its rapid progression, for this to be a possible.

Disease is characterised by a massive swelling of the affected muscle mass - although be aware that the lesion can be quite tiny and limited to a small area of cardiac muscle, the base of the tongue, or even an area of muscle within the gut wall and so not immediately obvious even if a detailed post-mortem examination is carried out. On palpation there is sometimes a classic feeling of popping bubblewrap.

Other species of clostridia target other sites within the body, although the outcome of disease, sudden death, is usually similar. *Clostridium novyi* type B causes an infectious necrotic hepatitis ('black disease') frequently associated with the damage caused by the migration of immature liver fluke through the liver parenchyma; *Clostridium novyi* type D (formerly referred to as *Clostridium*

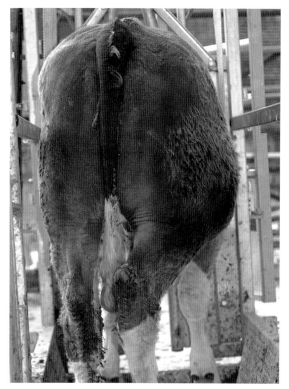

Unusual cases of blackleg, first because the animals are both alive, and both lived following treatment, and second because the Clostridial organism involved was not identified with certainty so these may be cases of malignant oedema or false blackleg, although the difference is academic!

haemolyticum) causes a condition known as Bacillary haemoglobinuria, where, if an affected animal is identified before death, it may be noticed passing dark or red urine – this may lead to a mistaken diagnosis of redwater fever. Clostridial organisms have also been defined as the cause of haemorrhagic abomasitis and enteritis in calves, and may have a role in the relatively recently described, and usually fatal, necrotic enteritis that typically affects suckler calves between two and six months of age.

Tetanus and Botulism

Tetanus and botulism are unusual amongst the clostridial diseases in that affected animals are usually identified before, rather than after death. Tetanus is caused by the neurotoxin

A calf affected with tetanus following infection around the rubber 'elastrator' ring used to castrate it. Note particularly the extended, stiff limbs, the head arched backwards over the shoulders (opisthotonos) and the raised tail-head.

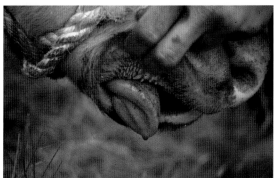

A Simmental x suckler cow affected with botulism, showing the classic signs of depression, flaccid paralysis, and in the second image, an inability to retract her tongue back into her mouth after it has been pulled out.

produced by *Clostridium tetani* following bacterial multiplication in the anaerobic environment provided by damaged tissues deep within penetrating wounds, including castration wounds; it is characterised by an extensor paralysis and muscular rigidity. Early in the course of this disease, or in less severe cases, infection is characterised by stiffness, muscle tremors and a reluctance to move, bloat (caused by an inability to eructate), protrusion of the third eyelids, rigidity of the ears, prominence of the tail-head, and the adoption of a wide-based, 'rocking-horse' stance.

The presentation of botulism is caused by the ingestion of the neurotoxins produced by *Clostridium botulinum* bacteria multiplying in the carcasses of dead birds and small rodents, classically the carcasses of dead chickens inadvertently spread on to grazing pastures when chicken-house waste is used as a fertiliser, but also poisoned rodents that have died within stores of conserved forage. In this case disease is characterised by a progressive flaccid paralysis causing ataxia, progressing to recumbency, with a classic sign being an inability to withdraw the tongue back into the mouth after it has been pulled out.

Treatment and Prevention: Clostridial Vaccines

Irrespective of the exact clostridial organism involved, treatment of affected animals that are identified before death is frequently unrewarding, and time-consuming in the cases of tetanus and botulism. Prevention, therefore, remains the only sensible option. A range of excellent clostridial vaccines exists, available at very reasonable cost, to protect cattle (and sheep) against the effects of one or more of these bacteria and the toxins they produce, and it remains a mystery why their use is not much more widespread within our national, and the global, cattle herd.

PROTOZOAL DISEASES

Neospora

Neospora caninum is a protozoal parasite remarkably similar to *Toxoplasma gondii* (a protozoal parasite of cats that is responsible for causing a significant number of sheep to abort each year), and was first described as causing an ascending hind-limb paralysis of puppies in 1984 (although it was not named until five years later). Now it is the most commonly recorded infectious cause of bovine abortion in the UK.

Neospora caninum lifecycle

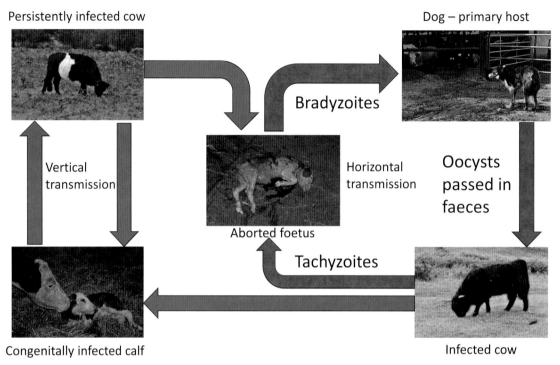

The lifecycle of *Neospora caninum* showing the more common 'vertical' route of parasite transmission from cow to calf, and the 'horizontal' route of transmission involving the primary host of the parasite, the dog, which infects adult cattle with oocysts passed in its faeces.

Dogs and other members of the genus *Canis* (wolves, coyotes, jackals and dingos) are the definitive hosts for the parasite. (Foxes, in the genus *Vulpes*, are not definitive hosts and are not thought to be significant in the epidemiology of the parasite.) Dogs are infected after ingesting bradyzoite tissue cysts in tissue from an intermediate host. Clinical signs are rare (except perhaps following congenital infection), but following sexual reproduction of the parasite in the dog's gut, large numbers of oocysts can be passed in the faeces, albeit usually only for a limited period of time.

Cattle are one of the intermediate hosts of the parasite, and can be infected either by ingesting sporulated oocysts, or congenitally when the tachyzoite form of the parasite passes from an infected dam across the placenta to infect the foetus, which it does with a high degree of efficiency; most live offspring of infected cows are born congenitally infected and remain so for life. (Cow-to-cow transmission is not a feature.) The outcome of infection in cattle depends on whether the challenge is exogenous from sporulated oocysts in the environment, or endogenous due to a recrudescence of a quiescent infection or a trans-placental infection of a foetus, when it occurs, particularly relating to the stage of pregnancy in the case of pregnant animals, and previous history.

Abortion, which typically occurs between five and seven months of gestation, is the likely outcome of foetal infection, which occurs between a few weeks after conception and mid-pregnancy. Foetal infection later during pregnancy more often results in the birth of a congenitally infected calf that itself, if it is female and retained as a breeding animal, is about three-and-a-half times more likely to abort than an uninfected herd contemporary.

Serological testing, ideally undertaken around the time of abortion or calving, can be used to identify infected animals, although a negative serological test result does not confirm the absence of infection. Similarly, a seropositive test result does not confirm Neospora as the cause of an abortion; this requires the use of histological techniques to demonstrate lesions typical of the parasite, ideally in foetal brain stem but in cardiac septum as an alternative.

Managing Neospora Infection in Cattle
Currently there are no effective treatments available to eliminate this parasite, nor any effective vaccines to protect against it in the UK. Herd-level control, therefore, relies first on removing and disposing of all potentially infected bovine material, including afterbirths, aborted foetuses and stillborn calves, so that these cannot be eaten by dogs, and by preventing dogs defaecating in cattle feed; and second, on identifying infected cows so that their female offspring are not retained as breeding replacements for the herd. (Bulls are not considered significant in the transmission of this parasite.)

Redwater Fever
Redwater fever is caused by the single-celled protozoal parasite *Babesia divergens*, which is transmitted by ticks, almost exclusively *Ixodes ricinus* in the UK. Clinical cases of the disease are seen, therefore, affecting susceptible animals at times of the year when ticks are active. All stages of the tick can transmit the organism.

Within the tick the parasite multiplies within the ovaries, infects the eggs, and then undergoes further multiplication within the larvae after hatching. Vertical transmission to subsequent generations of ticks is possible. Within the ticks the parasite migrates to the salivary glands, from where it can invade the mammalian host as the insect feeds.

Within the cow the parasites invade the red blood corpuscles (RBC) where they divide asexually, causing the RBCs to rupture, releasing the parasites to invade more RBCs. In addition to releasing the parasites, the ruptured RBCs also release their haemoglobin, accounting for the discoloration of the urine due to haemoglobinuria, from which this condition takes its name. Depending on the severity of the situation, which depends on the magnitude of the challenge and the level of immunity of the affected animal, the degree of discoloration seen affecting the urine will vary from almost unnoticeable through the classic deep red of port wine, to almost black. Affected animals will have a raised temperature, and depending on the extent of the condition, may become weak, pale, exhibit respiratory distress and even die, effectively of suffocation if insufficient intact RBCs remain to carry sufficient oxygen around the body.

Suspicion of the disease is often based on finding a dull and depressed animal with a high temperature on known high-risk pastures during tick season. Haemoglobinuria will heighten this suspicion, and confirmation of the cause can be obtained by demonstrating the presence of the parasites within RBCs when stained blood-films are viewed using a microscope.

BLOOD TRANSFUSION IN CATTLE

Blood transfusion is rarely performed in cattle practice, although when it is, it can be a life saver. It is indicated where there has been a significant haemorrhage, assuming that the source of the haemorrhage can be identified and addressed, or where haemolysis (as occurs in cases of redwater fever) causes the PCV (the volume of the blood made up of red blood corpuscles) to fall to below 12 or 15 per cent (it would normally be around 30 per cent in cattle), depending on the demeanour of the animal, but remaining above about 8 per cent. (The outcome of transfusion in cases where the PCV has fallen below 8 per cent is rarely positive.) In most cases where transfusion is carried out the aim will be to transfuse 5ltr of whole blood to allow time for other treatments and supportive therapy to take effect.

Blood groups in cattle are complex and rarely determined. They do have the potential to influence the success of transfusion, particularly because of the possibility of anaphylactic reactions. However, these are unlikely following a single transfusion, and because the indications for transfusion are invariably life threatening, the risk is usually considered worth taking and can be reduced by the administration of corticosteroids at the time of the transfusion.

Donor animal selection needs careful consideration. A quiet animal in good body condition and health from the same herd will facilitate collecting the blood for transfusion, ensure minimal adverse effect and risk of complication for the donor, and minimise the biosecurity risk to the recipient. Usually a barren cow or a fattening steer will be chosen. Adequate restraint for the donor animal will be required providing good access to the jugular vein in the neck. Sedation may facilitate blood collection but is not necessary in all cases.

The blood collected must be mixed continuously with an anticoagulant as it is collected in order to prevent clotting. The anticoagulant of choice in cattle is 3.8 per cent sodium citrate, which, for a 5ltr transfusion, can be pre-prepared by mixing 19g sodium citrate in a 500ml pack of saline for i/v administration.

After the donor animal has been adequately restrained, its head should be pulled securely to one side to expose the jugular groove, and the site from which the blood is to be obtained adequately prepared; a tourniquet (a bandage wound round the neck over a roll of bandage in the jugular groove, a clean calving rope or even a length of baler twine will suffice) should be placed around the neck below this site. The clinically superior method of collecting the blood for transfusion is to use a large-bore catheter and line to collect the blood directly into a sterile pack.

However, collecting 5ltr of blood from a donor animal is not a quick process, even using the largest bore catheters and lines available, and even if the catheter and line have been pre-rinsed with anticoagulant the process is frequently complicated by the formation of clots within the line before a sufficient volume of blood has been collected. For this reason, in practice the blood is often collected into a clean bucket containing the anticoagulant by making an incision between one and two inches long into the jugular vein after the administration of local anaesthetic. (It is surprising, even when using such a collection method, how often the blood will clot and stop flowing before the full 5ltr target volume has been collected!)

After sufficient volume of blood has been collected, the incision should be closed with a couple of mattress sutures; this task may be made easier if these have been pre-placed. The donor animal should then be kept quiet, offered a drink and good quality food before being returned to the herd.

After the required volume of blood has been collected, and added to an appropriate container for administration if it was collected in a bucket, it can be transfused into the patient, ideally through a giving set incorporating a filter to remove any micro-clots that have formed during the collection process, in as stress-free manner as possible. Additional treatment and supportive therapy should also be given at this time, again avoiding stressing the patient as much as possible; the lower their PCV, the smaller the degree of stress required to tip them over the edge, resulting in their loss. When successful, however, the difference that a transfusion can make can be almost miraculous: a cow that a few minutes ago was lying down and unable to lift her head may stand and start eating as if nothing has happened! Ongoing care and attention, and perhaps more treatment, will, however, still be required to maximise the probability of a successful outcome.

Treatment

The only licensed treatment for redwater fever currently available in the UK is imidocarb, which can also be used to prevent disease where animals with a predictably poor level of immunity are being introduced on to high-risk pastures during tick season (but beware the prolonged meat withdrawal period that is required following the use of this product, and the requirement to notify the APHA of its use). Oxytetracycline has also been recorded as having positive effects, and supportive treatments – including, in the most severe cases (where the PCV has dropped below perhaps 12 per cent or even into single figures), blood transfusion from another healthy animal – will aid the achievement of a successful treatment outcome (although this is far from certain, especially in the most severe cases).

Increase in Cases of Redwater Fever

Changes to the areas in which cases of redwater fever have historically been seen, and the time of year at which they are seen, may be occurring, possibly as a consequence of climate change and global warming. If this is the case, it can be expected that this expansion, both geographically and seasonally, will continue as tick distribution and their period of activity increases. Just because redwater fever 'has never affected animals here before' is no longer a reason to exclude it from a differential list now or into the future. (The same reasoning should also be applied to Anaplasmosis and Ehrlichiosis caused by Rickettsial organisms that are also transmitted by ticks and attack the RBCs and leucocytes of cattle – and other species – respectively causing pyrexia, immunosuppression and, potentially, significant mortality.)

HEALTH MONITORING AND DEFINING HEALTH STATUS

Designing a testing programme to define individual animal or herd health status and then to monitor this over a period of time is not easy.

Careful thought needs to be given to what is needed to be known, what you are looking for, and which test or tests to use. Do you, for example, need to know whether a particular infectious agent or disease is present within a herd, in which case testing should be concentrated on those animals most likely to be carrying the potential pathogen as only a single positive result is required to confirm its presence? Or do you require information about the prevalence of infection within a herd, in which case a statistically significant number of representative animals, depending on the herd size and profile and the epidemiology and pathogenesis of the agent of interest, should be sampled and tested? Or are you seeking reassurance about the absence of an infectious agent or disease from a herd, in which case it might be necessary to sample and test all the individuals in the herd and on multiple occasions over an extended period of time?

Consideration also needs to be given as to what type of test is used, and test performance.

Type of Test

Direct tests, bacterial culture, virus isolation, antigen detection or PCR (which detects fragments of pathogen genome) detect the presence of the potential pathogen directly. These may be useful when hunting for persistently infected animals within a herd as part of a BVD eradication programme, but may miss infected animals when the infectious agent has established a latent infection and hidden itself away, as is common with herpes virus infections for example.

In addition, direct testing technologies are unlikely to inform about previous exposure of the animal to the potential pathogen. For this, an indirect testing technology that detects the body's response to challenge and possible infection, usually in the form of specific antibodies, might provide more and more relevant information; varying levels of antibody response across all ages of animals within the herd might suggest a recent or on-going challenge and the continued presence of the pathogen within the herd, while the presence of antibodies in only the older animals in the herd, with the younger animals remaining seronegative, might suggest historic infection that has now been eliminated and the absence of on-going challenge.

Interpretation also requires a knowledge of the pathogenesis of the specific pathogen of interest. A positive antibody result in the case of BVD, for example, indicates the probability that the host's immune response has eliminated the infection; but in the case of IBR, a positive antibody result is suggestive of latent infection, with recrudescence and shedding of the virus being possible at any time.

Test Performance: Sensitivity and Specificity

The performance of all tests is defined by sensitivity and specificity, with an understanding of predictive value also being necessary to correctly interpret test results. Sensitivity refers to the ability of a test to correctly identify infected animals; it is a measure of false negative results that may be a direct consequence of test performance, or a consequence of the behaviour of the infectious organism within the animal. Furthermore the progression from infection to disease, and therefore the body's response to that infection, can be particularly slow – for example, in the case of most Mycobacterial infections, many of which may remain permanently quiescent and never progress to clinical disease, limiting changes for the test to detect.

Specificity indicates the ability of a test to correctly identify the presence of a pathogen: it is a measure of false positive results that may be seen when the animal has been exposed to, or infected with, an innocuous environmental organism very similar to the pathogen of interest, for example. Mycobacteria again provide a good example, with a plethora of environmental Mycobacterial organisms potentially resulting in false SICT test results - hence the use of the SICCT test, which introduces a comparative component to the testing technology to improve specificity when aiming to identify *M. bovis*-infected animals. Predictive value then adds weight to a particular test result: the higher the prevalence of infection and disease within the herd, the more likely it is that a positive test result will be correct, so the higher its predictive value.

Sensitivity and specificity can be altered by altering the interpretation of the test data, but only at the expense of the other: increasing the sensitivity of a test almost always decreases its specificity, and *vice versa*. Changing cut-offs to alter test performance can be a very valuable tool. If the aim is to eradicate infection, BVD for example, missing a single persistently infected animal because of a false negative result would be a disaster. In this situation test sensitivity will be paramount.

However, when screening a herd for the presence of an infectious agent or disease that is believed to be absent, false positive results may be more problematic, and the importance of test specificity is increased. Consideration needs to be given in such a case to the consequences of positive results: will animals that give a positive test result be culled from the herd, or simply managed differently?

Useful information and guidelines about health screening and defining health status can be found within the CHeCS (Cattle Health Certification Standards) technical document.

BIOSECURITY

Sampling and testing for exposure to potential pathogens and the presence of infectious disease only provides a snapshot of herd status at that point in time. More dynamic information can be obtained by serial screening over time. Even this, however, is of limited value without attention to biosecurity: without adequate biosecurity new challenges to health status may occur at any time, and the situation may change rapidly.

Biosecurity needs to address many wide-ranging possible sources of threat to the health status of the herd, including visitors to the farm (especially the vet!) and wildlife vectors of disease: the badger always springs to mind as a potential vector for the introduction of *M. bovis* and bTB into a herd. The greatest threat, however, to any established population of any species is always another individual of the same species – in this case another cow, with the two major threats being nose-to-nose contact (albeit that some infectious agents can be spread either in droplet form, the foot and mouth disease virus for example, or by insect vectors, bluetongue and Schmallenberg viruses, blown by the wind) between the animals in your herd and cattle of unknown health status on neighbouring holdings, and the introduction of purchased or hired, in the case of bulls, animals into your herd. Both, however, with a little thought and effort, should be relatively easy to manage.

Boundary Biosecurity

It is commonly quoted that boundary biosecurity requires the presence of robust and intact double fencing with a gap of 3m between the fences. This will certainly achieve the desired aim, but it is an expensive option, not least because of the amount of potentially productive land that can no longer be utilised, and so is often not done. However, it is not the only solution to the problem.

Temporary electric fencing alongside a more permanent fence while cattle are grazing a boundary field will achieve the same aim, as will a thick, well maintained hedge or a high, solid wall. More inventive solutions include shutting up your boundary

A barbed-wire fence and a well laid hedge with space between them sufficient to prevent nose-to-nose contact between cattle on either side of a farm boundary, aimed at improving biosecurity.

fields to cut for hay or silage when cattle are being grazed in neighbouring fields, or agreeing with your neighbour to stagger the presence of cattle in neighbouring fields (or even agreeing to monitor and eradicate specific infectious agents concurrently).

Quarantine

So often quarantine, if it is carried out at all, comprises a stay for two or three weeks in the dingiest, darkest shed at the furthest reaches of the farm, and if the incoming animals eat the forage that is thrown to them, then all must be well. Nothing could be further from the truth! Quarantine should be a proactive process, beginning before the animals destined to enter the herd even leave their farm of origin by seeking reassurances about, and evidence of, their health status and that of their herd of origin.

Once in quarantine, which should eliminate the possibility for any contact,

Electric 'badger' fencing constructed around the perimeter of the cattle buildings and the farmyard, aimed at improving wildlife biosecurity.

both direct and indirect, between the incoming animals and the established herd, the incoming animals should be critically inspected on a regular basis for any signs of ill-health or infectious or contagious disease. In addition, during the quarantine period samples should be collected and tested to confirm the health status of the incoming animals, and also, as far as is possible, the absence of risk to the health status of the established herd. In addition any prophylactic treatments for, for example, parasite control should be completed, and any necessary vaccines should be given, with primary courses completed before the animals are released from quarantine.

Perhaps one of the most debated questions about quarantine is how long it should last. The answer to this cannot be given in a specific and consistent number of days or weeks: it depends on the aim for the process. In most cases it should be a period long

enough for relevant testing to be carried out and the results acted upon, and for relevant prophylactic treatments to be given and vaccination courses completed – but specific situations may require a prolonged quarantine period. It is currently not possible, for example, to determine with any certainty the BVD status of an unborn foetus *in utero* within a BVD-seropositive dam, so quarantine should arguably continue until the calf has been born and tested. (Should pregnant, seropositive animals be considered for addition to the herd? This begs the question whether pregnancy diagnosis should be included in the quarantine testing protocol.) Furthermore the time between infection and the onset of clinical disease in the case of Mycobacterial disease can be so long that perhaps quarantine should be regarded as equivalent to the lifespan of the animal (certainly serial testing should continue for such a period of time)!

CHAPTER 9

PARASITES

Cattle and their parasites have evolved together, side by side, over many hundreds of thousands of years. The presence, therefore, of a low level of nematode parasitism should not be a surprise, and perhaps should even be expected, and should not, in fit and healthy, well-fed animals and with good pasture and grazing management, cause any great problem. Problems can, however, and do arise where poor pasture management, excessive stocking density and inadequate parasite control programmes, including an over-reliance on, and use of, anthelmintic products, result in increased challenge and poor immunity.

ENDOPARASITES

Gut Worms

A range of gastro-intestinal nematode, or roundworm, parasites infect cattle. The majority of these, despite having predilections for different parts of the gut, share a similar lifecycle: adult worms produce eggs that are passed in the faeces (so in most situations a faecal worm egg count can give a useful indication of the parasite burden present), these hatch on the pasture, and develop into the infectious, usually L3 larval stage of the parasite with the potential to reinfect grazing cattle following ingestion. After ingestion and a further series of moults, usually taking

about three weeks, the larvae develop into egg-laying adults, and the cycle begins again.

Of particular significance, because of the phenomenon of hypobiosis, is the abomasal parasite *Ostertagia ostertagi*. During the summer months this trichostrogyle worm behaves much like any other gut worm. Low levels of parasitism will compromise digestive and absorbative efficiency but will often go unnoticed, although poor growth rates may be reported. As parasite numbers, and therefore damage to the gut wall increase, subclinical disease will progress to clinical or Type I disease characterised by poor health and performance, scouring (which may be profuse), weight loss, and even death in the most extreme cases.

As the season progresses, however, and day length shortens, an increasing proportion of the ingested L3 larvae do not continue to develop to adulthood but burrow into the abomasal wall, where they remain encysted in a semi-dormant or 'hypobiotic' state until their synchronous emergence as day length increases in the spring. This can result in massive damage to the abomasal wall, causing a profuse, acute and not unusually fatal scouring in young, housed cattle, Type II disease, just before turn-out for their second grazing season. Such cases can catch the unwary by surprise: how can housed cattle be suffering a fatal worm infestation, especially

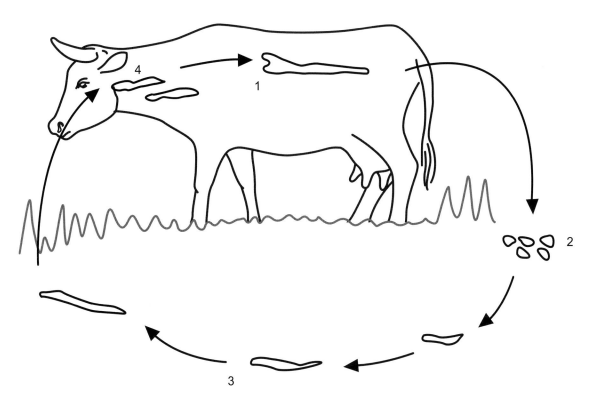

Generic nematode lifecycle: adult parasites in the cow's gut (1) produce eggs that are passed in the faeces (2). After hatching, the larvae develop until they reach the infective L3 stage (3). If they are then ingested, their development will continue within the cow through further larval stages (4), until they mature into an adult worm (1), which will begin producing eggs after about three weeks.

when faecal worm egg counts are negative (because it is a larval stage of the parasite that is causing the problem, rather than egg-laying adults)?

Prevention

Prevention is relatively straightforward in theory: manage young animals during their first grazing season to maximise the development of immunity, and treat them with an appropriate anthelmintic at housing – however, although some wormers kill worms, not all of them do. Some, Levamisole-based products for example, paralyse the parasites that they are targeting. This can work well for parasites within the gut lumen, which then relax their

grip on the gut wall and are passed in the faeces – but where the parasites are encysted within the gut wall, even if paralysed by the product administered, they remain within their cyst until the paralysis wears off, so the course of the disease remains unchanged!

Health and productivity can therefore be optimised by appropriate anthelmintic treatment, but beware: not all wormers, as hinted at above, do the same thing, and the over-use of wormers, particularly containing active ingredients from the same group, will almost inevitably result in the development of resistance amongst the population of parasites being treated. All populations of worms will contain

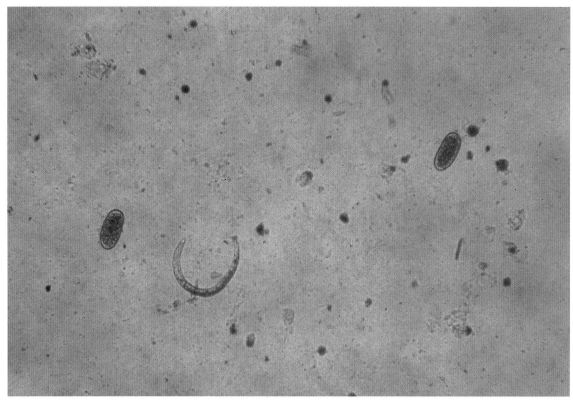

Gastro-intestinal nematode eggs, along with a lungworm larva, in a faecal floatation preparation under microscopy; faecal worm egg monitoring is an essential part of monitoring and managing gut-worm burdens.

some parasites that are resistant to one or more of the commonly used anthelmintic groups.

Although the majority of any gastro-intestinal nematode parasite population will be *in refugia* on the pasture, and therefore not exposed to any anthelmintic product administered to the animals grazing that pasture, when an animal is treated, only worms resistant to the product used will persist within the animal, producing eggs that are then added to the *in refugia* population of the parasite in the animals' dung. Over time the proportion of the *in refugia* population of the parasite that is resistant to the anthelmintic being used will therefore increase, and the efficiency of

the anthelmintic product will decline until it becomes almost useless. It is therefore of vital importance to treat according to 'COWS' (control of worms sustainably) principles: only treat animals at pasture when necessary, and ensure accurate dosing when they are treated.

Worming at housing is subject to slightly different considerations: any worm eggs that are shed by the now housed cattle will no longer be viable when they are eventually spread on the land in manure or slurry the following spring, and it is necessary to worm in particular young animals being housed after their first grazing season in order to prevent Type II ostertagiasis. (It also means that you are

THE DIFFERENT CATEGORIES OF WORMER

The various wormers available with which to dose suckler cows – or, more likely, their calves (and sheep and other animals) – are categorised according to their mode of action. This is important in the development of resistance amongst parasite populations to the products available for use, and requires careful consideration before choosing which anthelmintic to use.

Group 1: Benzimidazoles (white wormers): The large number of products available in this class disrupt cell structure, killing the target parasites, making them suitable to treat encysted hypobiotic parasite larvae as well as adult parasites. They are only available as oral formulations (drenches and boluses).

Group 2: Levamisoles (yellow wormers): Levamisole-based anthelmintics paralyse nematode parasites, causing them to lose their grip inside the body: this allows gut worms to be passed in the faeces, and lungworms to be coughed up. However, hypobiotic stages of gut worms will recover from the treatment whilst still encysted within the gut wall, and so this class of anthelmintic is not suitable for the treatment of young, housed cattle during the winter. Injectable, oral and pour-on preparations are available.

Group 3: Macrocyclic lactones (clear wormers): Macrocyclic lactone anthelmintics (the 'mectin' wormers) are GABA agonists and block the passage of nerve impulse within a wide range of nematode and invertebrate parasites; they are therefore widely used in the control of both endo- and ectoparasites. However, they also have a possibly significant effect on beneficial invertebrate species, particularly dung beetles (which compromises the degradation of cow-pats), in the environment. Although available as injectable and oral preparations, the wide range of cheap, convenient pour-on preparations available has led to their widespread use, which has been followed, unsurprisingly, by increasing reports of resistance amongst parasite populations.

Group 4: Amino-acetonitrile derivatives (orange wormers): The only anthelmintic available in this category is Monepantel. It is only licensed for use in sheep in the UK.

Group 5: Spiroindoles (purple wormers): The only anthelmintic available in this category is the combination product 'Startect' (a combination of derquantel with the macrocyclic lactone abamectin). It is only licensed for use in sheep in the UK.

not feeding a parasite burden all winter with expensive conserved forage or, worse, concentrate feed!)

Worming at turnout, although this undoubtedly provides a convenient opportunity to treat the animals, should not be required if they have been adequately treated at housing, and adult cattle fed well and grazing well managed pastures should not routinely require anthelmintic treatment.

Tapeworms

Tapeworms are flatworm cestode parasites that have a dual host lifecycle. Adults, which comprise multiple segments in linear array (ribbon-like), usually live in the gut of the primary host species, while the immature form of the parasite is found, usually in an encysted form, in the intermediate host.

Cattle, along with sheep and goats, are primary hosts for *Moniezia* tapeworms, which have forage mites as their intermediate hosts. Infestation is frequently obvious because of the passage of segments of the parasite, either individually or in ribbons, in the faeces; but although unsightly, this is considered of little clinical significance (although where particularly massive burdens are present a physical blockage of the gut may cause problems).

Cattle can, however, act as the intermediate host for some of the Taenid tapeworms, which predominantly have the dog as their primary host, *Taenia multiceps* and *Taenia hydatigena* for example (although man is the primary host of *Taenia saginata*). *Echinococcus granulosus* has the fox as a primary host, but its immature stage can develop in a range of intermediate hosts, including man, causing hydatid disease. This is why it is so important to prevent the consumption of dead animals by farm dogs, and to make sure that farm dogs are regularly and adequately wormed with an anthelmintic active against tapeworms - praziquantel for example.

Clinical signs associated with the immature stages of these parasites in cattle are rare, although they do occasionally occur, depending on the site where the tapeworm cyst develops (gid in sheep is a consequence of the development of the cystic larval stage of *Taenia multiceps*, *Coenurus cerebralis*, forming a space-occupying lesion within the brain). More common is their detection at meat inspection when the identification of multiple cysts, *Cysticercus tenuicollis*, the larval stage of *Taenia hydatigena*, may result in organ or even, in extreme cases, carcass rejection and economic loss.

LUNGWORM

The lifecycle of the lungworm parasite, *Dictyocaulus viviparus*, is similar to, but with important differences from, the lifecycles of

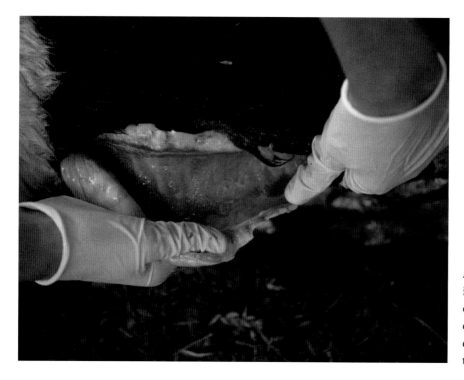

Adult lungworm in the trachea, confirming the cause of the respiratory distress that killed this cow.

A Hereford x cow suffering from clinical lungworm.

the various species of gut worm. The adult parasites live, permanently coupled, in the airways in the lungs, explaining the clinical presentation of infection, coughing in more mild cases progressing to respiratory distress and even death in the most severe cases.

Eggs are laid by the female worms, are coughed up into the oro-pharynx, are swallowed and hatch during their passage through the gut. Larvae are passed in the faeces on to the pasture, where they spread from the dungpat by climbing up *Pilobolus* fungi to be dispersed along with the fruiting bodies of the fungus. After moulting twice to reach the L3 larval stage the parasite becomes infectious; if it is ingested on blades of grass it will migrate through the gut wall and to the lungs, before further moults to reach the adult stage back in the lungs, when the cycle starts again.

A degree of immunity can be developed and then maintained by cattle as a consequence of low level, continued exposure to the parasite (or vaccination with irradiated L3 larvae followed by a trickle challenge from the pasture), but this is not particularly robust and will be overcome during periods of overwhelming challenge – typically during the autumn following rain after a long, hot, dry spell during the summer. (During hot, dry weather the lungworm larvae will migrate down the leaves and stalks of the grass plants seeking out the damper environment amongst the roots where they will not usually be ingested by animals grazing the pasture. When it rains, however, they reverse their direction of travel, migrating up the leaves and stalks of the now damp grass plants to present a massive and synchronised challenge to the animals on the grazing, which will be exacerbated by the overuse of anthelmintics in the herd, limiting the preceding challenge and therefore immunity.)

Treatment

When clinical disease strikes, a conundrum presents: if there is one thing worse than a lungful of live lungworm it is a lungful of dead lungworm – and yet affected animals have to be treated. It makes sense, therefore, to use an anthelmintic preparation that paralyses rather than kills the parasite so that the adult worms relax their attachment to the airway walls and can be coughed up, rather than dying and remaining *in situ*. Anti-inflammatories and antibiotics also make sense as part of the treatment plan – but even so, some of the worst affected

Liver fluke life cycle: adult fluke living in the bile ducts in the liver (1) produce eggs, which travel in the bile into the gut lumen and are then passed in the faeces (2). After two to four weeks the eggs hatch releasing miracidia (3), which parasitise mud snails, *Galba truncatula* (4). After a further six weeks or so, cercaria (5) emerge from the snails, move up the grass and encyst as metacercariae (6), which can remain viable for many months. If eaten by a cow (or sheep, or other susceptible animal) the metacercariae develop into immature liver fluke; over the next twelve weeks these migrate through the cow's body and liver before maturing into the adult stage of the parasite in the bile ducts (1).

animals will get worse and may even die following treatment.

Liver Fluke

Liver fluke, *Fasciola hepatica*, is another nematode parasite of cattle, but this time a trematode, or a flatworm, rather than a roundworm. Their lifecycle requires the involvement of an intermediate host, a mud snail, *Galba truncatula*, and is significantly more complex than that of the nematode parasites of cattle.

Adult liver fluke live within the biliary system within the liver, intermittently producing eggs, which pass into the gall bladder and then the gut via the bile duct. The eggs can be detected in faecal samples, in a similar process to nematode eggs, although a different flotation technique is required. They are passed on to the

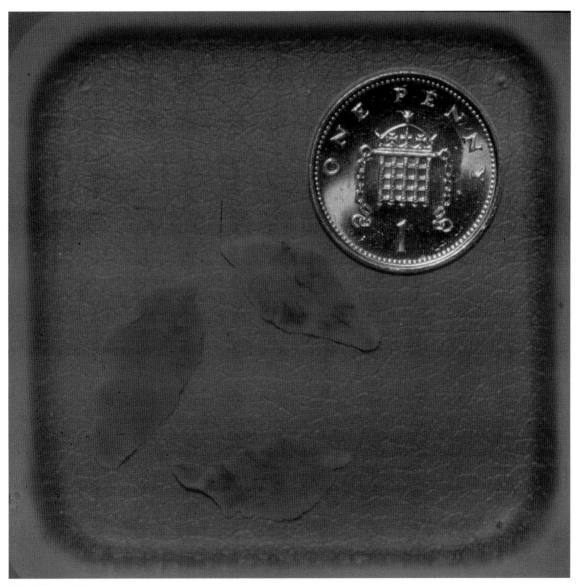

Adult liver fluke.

A suckler herd grazing high-risk pastures for liver fluke.

pasture in the faeces where, after just over a week, depending on climatic conditions and temperature, they release a motile miracidium: in order to remain viable and for the lifecycle to continue, these miracidia need to infect a snail within just a few hours.

Within the snail the miracidia continue their development over a further period of weeks or months (again depending on climatic conditions and temperature) until they are shed as cercaria: these move up the leaves and stalks of the grass plants in the pasture and attach themselves in an encysted form, developing into infective metacercaria. Metacercaria that are ingested along with the grass by grazing cattle emerge from their cyst within the gut, migrate through the wall of the small intestine and then through the peritoneum to the

liver: here the immature fluke penetrate the capsule and migrate through the liver parenchyma to the bile ducts, where, after ten or twelve weeks, they mature into adults – and the cycle begins again.

With such a complex lifecycle it will probably come as no surprise that the clinical signs caused by liver fluke can vary widely in both the time of the year that they present and their severity, with both being highly dependent on climate and temperature: not only is the development of the various stages of the parasite dependent on damp, warm conditions, but also the lifecycle of the parasite cannot be completed without the intermediate host, the snail, which also requires damp, warm conditions. Ambient temperatures consistently above 10°C are required for both the development of the parasite and for the snails to be active and to reproduce, so challenge to cattle during the winter in the northern hemisphere is likely to be minimal. Furthermore during the summer, when the weather may be hot and dry, the snails will retreat down into the soil, seeking out a damp environment, and so again, challenge to cattle is likely to be minimal. The major period of challenge in the northern hemisphere will therefore usually be during the warm, damp days of autumn when snail populations are maximal.

Depending on the level of challenge, acute disease may be seen (although this is perhaps more common in sheep than in cattle, partly as a consequence of the smaller size of their liver) as the immature fluke migrate through the liver parenchyma during the late summer and autumn. In cattle, however, chronic disease (which may be sub-clinical and often confused with an inadequate plane of nutrition) is caused by the presence of adult liver fluke within the bile ducts, and presents with a range of signs, from a cow that is 'under the weather' through to anaemia and hypo-albuminaemia, manifesting as severe weight loss and dependent oedema, which may be particularly noticeable under the jaw (bottle-jaw). Chronic disease probably presents more commonly during the late winter and early spring.

Prevention
The development of immunity to liver fluke is minimal, so prevention of disease caused by the parasite depends on good pasture management (and keeping lots of ducks to eat the snails!). Known high-risk pastures should not be grazed, if this is possible, at high-risk times of the year, and animals that can be predicted to have been exposed to the parasite should be treated with an appropriate product at an appropriate time.

This can be challenging, because not all flukicides are effective at killing all stages of the parasite. Some, albendazole and oxyclozanide for example, are only active against the adult stages of the parasite living within the bile ducts, and so, whilst useful to treat cases of chronic disease, are of no use to treat animals suffering from acute disease during the autumn. The only products available that target immature liver fluke down to a week or two of age are those containing triclabendazole. Such products make an ideal fluke treatment during the autumn, particularly at housing (or ideally two weeks after housing), but with the caveat that widespread and repeated use of these products year after year is reportedly resulting in increasing parasite resistance.

Nitroxynil provides an alternative treatment that will target adult and late immature, but not early immature stages of the parasite.

Rumen Fluke
Although of a different genus to *Fasciola hepatica*, the liver fluke, rumen fluke, from the genus *Paramphistomum*, share a similar lifecycle, although the intermediate hosts in this case are water snails, and the immature fluke develop and feed within the duodenum before migrating into the rumen where the adult parasites are found.

Where disease is seen, it is usually in young cattle (adult cattle, unlike the situation with liver fluke, will develop a level of immunity to the parasite, and even quite large burdens of the adult parasite within the rumen are well tolerated), often after wet periods as flood water lying over pastures subsides. It is a consequence of the immature fluke feeding within the duodenum and damaging the mucosa that causes an enteritis, haemorrhage and ulceration, characterised by weight loss and a bloody and sometimes profuse diarrhoea. Mortality may be high where challenge has been severe.

Treatment

If treatment options are limited in the case of liver fluke, they are even more so in the case of rumen fluke: oxyclozanide, which is active against the adult parasites, is one of the very limited number of effective molecules available.

ECTOPARASITES

Ticks

Although there is a range of both 'hard' and 'soft' ticks recognised across the globe, the predominant tick parasite of cattle recognised in Great Britain is the Ixioid

Adult rumen fluke.

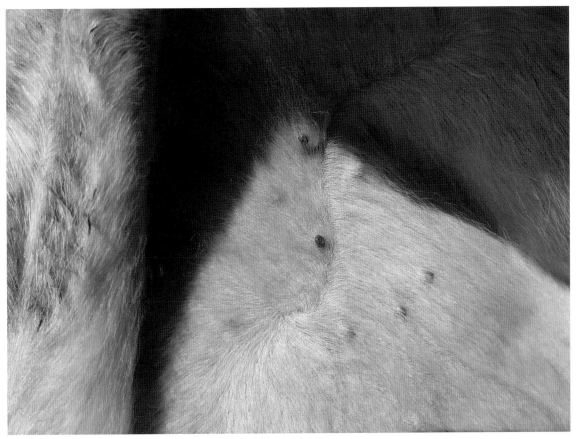

Ticks feeding on the inner aspect of the upper hind leg of a heifer being treated for redwater fever.

tick, *Ixodes ricinus* (although this situation may well change as a consequence of global warming and climate change). *Ixodes ricinus* has a widespread distribution around the world, including not only Europe but also the North American, Australian and African continents. It is found in Britain inhabiting areas of rough grazing, around field margins and on unimproved pastures including scrubby areas and heath and moorland.

The Ixioid tick is not limited to parasitising cattle, but will also feed from a wide range of wild and domesticated animals, and man, and has a prolonged lifecycle that requires a three-year period to complete, although the parasite only feeds for a few days during each of these years.

Historically in Great Britain this feeding period has had a seasonal pattern, with the peak periods for tick activity being during the spring and autumn. However, again perhaps as a consequence of global warming and climate change, there is evidence emerging that this may be changing, with tick activity being seen earlier in the spring and persisting later in the autumn than used to be the case.

The larvae hatch from eggs laid in the undergrowth and measure less than a millimetre long; they will attach themselves the following year to a passing mammalian host, attracted by the host's body warmth. They will feed for about a week before dropping off and moulting to become a nymph, measuring up to two millimetres long. The following year the nymphs will attach to another host, feed, drop off and moult to become an adult tick. During their third year the adult ticks will attach to a further host to mate and feed, after which the females may measure up to a centimetre long (although the males remain only a fraction of this size). Males may mate several times, but females mate only once, after which, and following feeding, which may last for up to two weeks, they drop off the host to lay their eggs and die.

Although tick bites can become infected, and where tick burdens are enormous they may cause anaemia, ticks themselves are not usually regarded as posing any significant pathogenic problem. Where they do have significance, however, is as a vector for other infectious organisms that pass from the tick in its saliva into the host during feeding. In humans, dogs and horses, Lyme disease, caused by the bacterium *Borrelia burgdorferi*, may be a significant threat, but in cattle redwater fever, caused by the protozoal parasite *Babesia divergens*, is a more significant issue, especially where animals with no immunity are exposed to ticks and the parasite for the first time.

Treatment
Although tick control in the tropics may still involve dipping cattle on a regular basis, topical acaracides are now more commonly used in Europe, along with avoiding grazing high-risk pastures during 'tick season'.

Mites
Many species of mite, including *Chorioptes*, *Psoroptes* and *Sarcoptes*, parasitise cattle causing mange. Whilst Chorioptic mange usually results in minimal irritation, with lesions commonly seen on either side of the tail-head of housed cattle, both Sarcoptic and particularly Psoroptic mange can cause intense irritation (*Psoroptes ovis*, which also parasitises cattle, is the cause of sheep scab, which can cause significant welfare issues and financial loss): this leads to affected animals spending significant amounts of time rubbing and scratching, which causes food intake and therefore performance to drop, and sometimes results in significant welfare issues.

Treatment
The treatment of choice will depend on the species of mite present, and whether it remains on the surface of the skin (*Chorioptes* and *Psoroptes*) or burrows into it (*Sarcoptes*): this can only be determined with certainty by examining the parasites under magnification. An additional challenge also now exists: there have been recent reports of macrocyclic lactone-resistant strains of *Psoroptes ovis* parasitising cattle in the UK.

Lice
Both biting (predominantly *Damalinia*) and sucking (*Haematopinus*, *Linognathus* and *Solenopotes*) lice can be found parasitising cattle. Numbers generally build up on housed cattle during the winter when the proximity of the animals facilitates the spread of the parasites, and the animals' winter coats promote their reproduction. Clinical signs are limited in most infestations, although the irritation caused by the more severe infestations, particularly of *Damalinia*, can result in excessive rubbing and scratching, and heavy infestations of sucking lice can result in anaemia, affecting performance, especially in young calves. However, rather than actually causing debility, heavy louse infestations are more frequently an indicator of debility because of some other cause.

Flies

Although a variety of Tabanid, Muscid and other species of fly feed on and around cattle, most flies, since the eradication of warble flies from the UK, are regarded as a nuisance rather than a serious threat to health. They may, however, cause irritation resulting in modifications to behaviour (food intake and therefore productivity may drop if a significant part of the animal's 'time budget' is involved in avoidance behaviour), or they may be involved in disease transmission (particularly mastitis and ocular infections), and on rare occasions, myiasis may be seen complicating neglected wounds or further debilitating diarrhoeic animals where there is faecal staining of their hind-quarters and tail-head.

Midges

Midges, of which there are very many different species, are generally regarded as an irritant rather than a serious direct threat to the health of cattle. They are, however, important as vectors of other infectious agents, including the blue tongue and Schmallenberg viruses.

Ectoparasite Control

Organophosphates have historically provided the cornerstone of ectoparasite control in cattle (and sheep), and their systematic use has successfully resulted in the eradication of warble flies from the UK; however, they are now only infrequently used because of their environmental consequences and their possible effect on human health. Their use has largely been replaced in recent times with synthetic permethrin-based products, frequently formulated as pour-on preparations applied along the backs of the animals. However, the environmental consequences of these products are now coming under increasing scrutiny, and attention is therefore increasingly being focused on natural methods of control; these include, for flies, the use of parasitic wasp larvae in areas such as muck heaps, for example, where flies are known to breed (although great care also needs to be taken to ensure an absence of currently unrecognised and unintended environmental consequences of these interventions!).

Pour-on permethrin-based products are not, however, a universally successful treatment or preventative measure for all ectoparasites of cattle, having little or no activity against some species of mites and biting lice. Where these are causing problems, treatment with a parenteral macrocyclic lactone-based product (ivermectin-based, doramectin-based and moxidectin-based products) is often advised, although this may need to be repeated depending on the persistency of the product chosen. Furthermore, reports of resistant strains of Psoroptic mites, resulting in welfare concerns due to the intense pruritis caused by the parasite, may limit treatment options.

CHAPTER 10

MISCELLANEOUS CONDITIONS

There are many other conditions that might affect the health status, welfare and productivity of beef suckler cattle on an individual basis, which are of little significance to the herd overall. For the individual affected they are, however, significant. Discussion here has been limited to some of the more common eye and skin conditions, causes of lameness and problems affecting the udder and urinary tract.

EYE CONDITIONS

There are two major infectious conditions that can cause eye problems in cattle: infectious bovine kerato-conjunctivitis, or New Forest eye; and bovine iritis, or silage eye.

New Forest Eye

New Forest eye usually begins as a relatively superficial issue caused by *Moraxella bovis*, which may initially only be apparent because of closed eyelids and tear-staining running down the face. In such cases, topical treatment with an antibiotic-containing eye ointment daily (or more frequently if possible) for two or three days may be sufficient to resolve the problem. If treatment is delayed, however, the condition can progress rapidly to the point that even aggressive parenteral treatment may be unsuccessful, and

surgery may be required to remove the eye. Nuisance flies and close contact between animals when eating concentrate feed from troughs have been implicated in the spread of the infectious agent between animals, so ensuring sufficient trough space and effective fly control will both help to reduce the number of cases.

Silage Eye

Silage eye is caused by listeria, often from mouldy or poorly made big-bale silage. Feeding silage in round feeders is a risk factor as animals eating on opposite sides of the feeder will often shake particles from

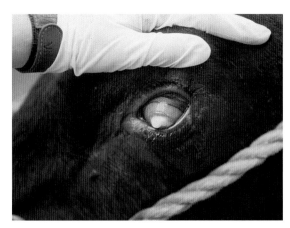

End-stage New Forest eye showing rupture of the cornea and protrusion of the contents of the globe. Although a persisting defect in such cases is inevitable, healing can be remarkably good following appropriate treatment, which should always include pain management.

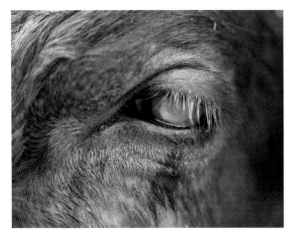

Bovine iritis, often referred to as 'silage eye', typically caused by listeria.

the silage into each other's faces and eyes as they are eating; feeding the silage at low-level rolled out on the ground will reduce the risk. The initial pathology includes an iritis within the eye, which can be intensely painful so NSAIDs should be included in the treatment plan. Parenteral antibiotic therapy is indicated.

'Cancer' Eye

Cancer eye, or ocular squamous cell carcinoma, is another eye problem not uncommonly encountered in suckler cows, especially those with a white face or white patches around their eyes; the Hereford breed is particularly susceptible, thanks to the effect of bright sunlight on these relatively unprotected areas of skin. Tumours can affect a variety of sites including the globe (eyeball) itself, the nictitating membrane (third eyelid), the eyelids or surrounding tissues.

Treatment options are effectively limited to surgery if the tumour can be removed in its entirety, which may be possible if it is sited on the globe or the nictitating membrane. Recurrence is common, however, so affected animals should generally be culled.

LAMENESS

Although many of the conditions that cause lameness in dairy cattle can also be found affecting beef cattle, incidence and prevalence are usually much reduced. In addition, the adage that 95 per cent of lameness affects the hind limbs, 95 per cent of that is due to a foot problem, and

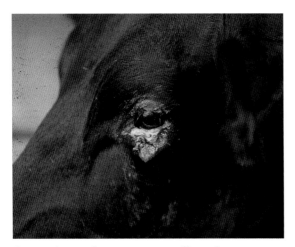

An erosive ocular squamous cell carcinoma affecting the lower lid of, somewhat unusually, a dark-skinned animal.

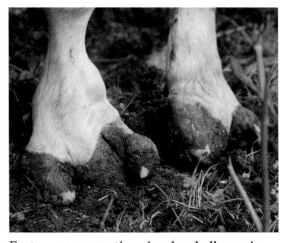

Foot care can sometimes involve challenges in the beef herd!

CATTLE FOOT TRIMMING USING THE FIVE-STEP 'DUTCH' METHOD

The five-step or 'Dutch' method of cattle foot trimming was formalised by Toussaint Raven, a professor at the veterinary school in Utrecht, to provide a structured approach to routine cattle foot trimming and the treatment of lame cows, and deliver a predictable outcome. Although the technique was principally aimed at dairy cows, it is just as applicable to suckler cows, although the issues encountered may differ.

Before any foot trimming is attempted it is important to have the animal safely (for both the animal and the trimmer) and securely restrained in a robust crush with a belly-band to help support the weight of the animal, and secure foot-trimming blocks to which each foot can be tied when it has been lifted. It is also important that the trimmer has the appropriate personal protective kit:

- wellies
- full waterproofs: close proximity to the rear end of the cow is required to trim the hind feet adequately, and if this is not achieved for fear of faeces running down your back you will not do a good job!
- wrist guards: it is surprising how much damage to soft human skin can result from the unavoidable impacts between wrist and the horn of the cow's foot you are trimming if protection is not worn.
- well maintained and sharp equipment, particularly hoof nippers (double-action hoof shears are not recommended) and knives (both right- and left-handed, irrespective of which hand the knives will be held in); and searchers rather than wider-bladed knives are recommended (double-edged knives should be avoided) if a good job is to be achieved. (Power tools should be avoided: although they can make the job quicker and easier, in inexpert hands they can also cause considerable damage. They should remain the preserve of professional cattle foot trimmers.)

As the name of the process suggests, the task is divided into five steps, with the first three aimed at restoring balance and normal weight-bearing to the foot – 'functional trimming' – and the final two aimed at addressing any lesions identified, and relieving weight-bearing to improve welfare and facilitate healing – 'curative trimming'. It is important that lesions are ignored during 'functional' trimming to ensure that correct foot balance is achieved before the lesions are addressed; indeed, it is surprising how many apparent lesions are insignificant and will disappear during the steps of 'functional' trimming – if these are addressed before the steps of functional trimming are carried out it might become impossible to balance the foot correctly.

FUNCTIONAL TRIMMING

Step 1
Starting with the more normal claw of the foot lifted (this will usually be the medial claw if it is a hind foot being trimmed and a lateral claw if it is a front foot although pathology may influence and alter this) the toe should be trimmed to length and the sole pared so that it is flat and bears weight evenly across its entire width.

Every cow, of course, is an individual and they all vary, so some toes will need to be kept longer and some should be trimmed shorter to achieve the optimal foot angle. Length is measured down the dorsal wall of the claw from the coronary band to the weight bearing surface where the wall comes into contact with the ground. A useful guide to length can be obtained by cupping the free hand around the lifted foot with the little finger placed along the coronary band and then using the width of the palm across the knuckles to direct how far to remove horn from the toe; there should be no reason for drawing blood!

When trimming the sole flat, in order to achieve the correct foot angle it will, almost without exception, be necessary to remove more horn from the outside edge of the foot compared to the inside edge and more horn from the toe than is removed from the heel; the appearance of the white line, the anatomical feature that occurs where the horn of the wall and the sole join, should inform that sufficient horn has been removed from the sole which should remain thick enough to protect the sensitive internal structures of the foot from the normal concussive forces encountered during locomotion, including on rough and uneven ground.

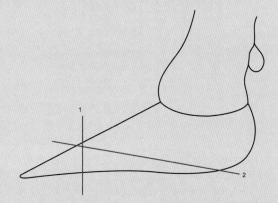

Steps 1 and 2 of functional trimming.

Step 2

Using the freshly trimmed, more normal claw as a template, trim the more abnormal claw to the same length and its sole flat and level with that of the more normal claw to ensure even weight bearing across both claws when the foot is put to the ground.

Step 3

With knives held vertically to the plane of the sole and inserted into the interdigital space, an elliptical 'model' is made relieving weight-bearing from the inner part of the sole of each claw and the medial wall adjacent to the interdigital space. This facilitates self-cleaning and aids ventilation of the interdigital space, reducing the prevalence particularly of soft tissue infections, interdigital phlegmon (foul-in-the-foot or lour) for example.

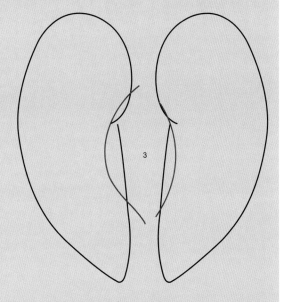

Step 3 of functional trimming.

CURATIVE TRIMMING

Step 4

This step involves removing all under-run horn. In dairy cows this frequently involves cracks in the horn around the bulbs of the heel, which result in a micro-environment deep within the cracks that allows infection and necrotic processes to become established. The damaged horn can be easily removed with a sharp knife by inserting the knife to the full depth of the most proximal crack (the crack nearest the coronary band at the bulb of the heel) and then removing the imperfect horn by moving the knife distally (away from the leg) whist rotating the wrist to preserve heel-horn height. Sole ulcers are another more common problem in dairy cows than in suckler cows although they can, and do, affect suckler cattle. The horn surrounding the ulcer is also frequently under-run, particularly in neglected or chronic cases, and also needs to be removed.

White line lesions need to be explored by removing an elliptical segment of wall horn to allow infection to drain without creating any pockets, particularly in the sole of the foot, which can become packed with mud and faeces that can then harden (ancient people used to build houses of such a mixture – this was the 'daub' in 'wattle and daub' – because of how hard it became when it dried!), sealing infection in.

Foreign body penetrations, including those caused by sharp stones, fragments of flint, discarded nails – particularly 'clout' nails with their broad head – screws and other shards of metallic debris, often cause more problems in suckler cows than in dairy cows. In such situations, although the damage visible on the outside of the foot may appear minor, there may be considerable damage within the foot; a nail penetration, for example, may only leave a relatively small discernible visible defect, but within the foot the nail may have penetrated and introduced infection into the digital cushion, the tendon sheaths and the distal inter-phalangeal, or coffin, joint – and if it has impacted the pedal bone the bone itself may be fractured. In these cases the degree of under-running of the horn of the sole may be considerable, and in such cases the best course of action might be to seek veterinary advice.

Step 5

This step is aimed at relieving weight-bearing from lesions to allow them to heal. This may be possible by trimming more horn away, for example by removing heel height across the entire width of the affected claw to relieve weight-bearing from a sole ulcer. It can also be achieved by fixing a block to the sound claw – but it is important that if this is done it is regarded as the start of the treatment rather than the whole treatment, and that the foot is re-examined after an appropriate period of time. It is also important to administer NSAIDs to any cow that is deemed to need a block. Furthermore a block will only work to relieve weight-bearing from the affected claw if the treated cow is then kept on firm footing: if she is kept in a deeply bedded hospital pen the block will simply sink into the bedding and fail to perform its function adequately.

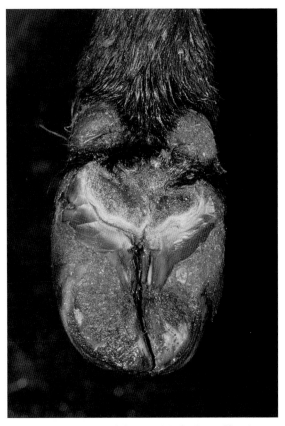

An extensive digital dermatitis lesion affecting the distal hind limb of a beef fattening animal.

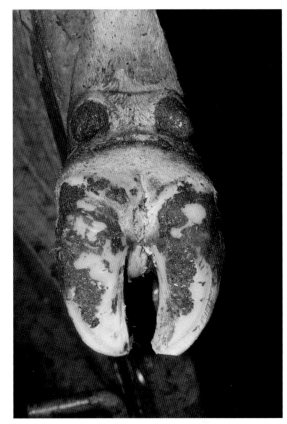

The typical interdigital swelling and splitting seen in cases of foul-in-the-foot.

95 per cent of *that* is because of a problem localised to the outer claw, may not hold quite so true in the beef sector as it does in the dairy sector, perhaps because of the much reduced exposure of beef cows to the concrete that dairy cows are exposed to, in most situations, every day. Claw-horn lesions associated with the erroneously named laminitis complex, including sole ulcers that are a common cause of lameness amongst dairy cattle, are not as frequently seen affecting beef cattle. Digital dermatitis caused by Treponeme bacteria, possibly the most commonly encountered infectious cause of lameness affecting dairy cattle, is not encountered to the same degree in beef cattle (although it does still occur).

Interdigital Phlegmon or Foul-in-the-Foot

Perhaps the most commonly encountered infectious cause of lameness encountered in beef cattle is interdigital phlegmon or foul-in-the-foot (or lour, or looe, depending on your local dialect!). This occurs when bacteria, including *Fusobacterium necrophorum*, invade small defects in the skin, often between the claws, and infect the surrounding soft tissue, causing sometimes significant swelling. Cases are often associated with poorly maintained, wet and unhygienic under-foot areas in

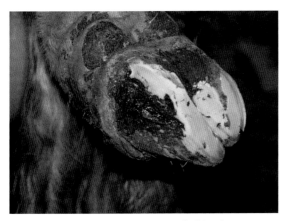

An infected vertical fissure, probably the result of a white line penetration but possibly a consequence of damage at the coronary band, resulting in defective production of the horn of the wall.

gateways or around water troughs. Treatment with one of a wide range of antibiotics coupled, of course, with a NSAID, will usually result in a rapid and complete resolution.

Vertical Fissures

Vertical fissures can present a management challenge. Although there is an environmental element to their aetiology there is also a genetic component, so perhaps the most successful approach to management is to 'breed them out of the herd'. Ensuring adequate vitamin and mineral nutrition, especially biotin and zinc, to promote healthy horn production, may help reduce the prevalence of true 'sand-cracks' (fissures that extend upwards from the bottom of the foot), although what is being fed now will not influence wall horn integrity for several months because of the time lag between the production of new horn at the coronary band and it coming into contact with the ground. Fissures arising as a result of damage to the horn-producing tissues at the coronary band and extending down the hoof wall are unlikely ever to heal.

A foreign body (in this case a shard of metal) has penetrated the sole of this breeding bull – performance will obviously be compromised (and may never be regained).

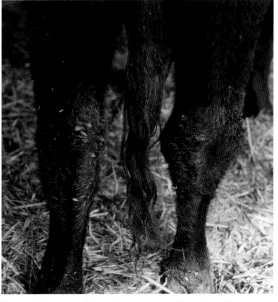

'Spavin' caused by the over-production of joint fluid following traumatic damage to the ligaments and menisci of the right hock of an Angus bull.

Damage Caused by Foreign Bodies

Whilst not common, trauma and damage caused by foreign bodies – flints, nails and wire, for example – may account for a greater proportion of the lameness cases that affect beef cattle compared to dairy cattle. Penetrating wounds affecting the sole of the foot deserve particular attention, because although the defect that is visible may appear minor, the damage within the foot, which cannot be seen, may be extensive. Treatment to remove all under-run horn and address any infection that may be present should not be delayed if a case of deep digital sepsis is to be avoided.

Traumatic Lameness

Breeding bulls, when kept in multiple bull groups with the cows during the serving period, present a particular challenge with respect to traumatic lameness. Fighting between the bulls often results in damage to the hock, and particularly to the ligaments and menisci involved in the stifle. While rest and treatment with NSAIDs may help and apparently resolve such cases, healing is rarely sufficiently robust to prevent recurrence when the bull is used again. In such cases, assuming recovery is achieved and the bull becomes able to walk relatively normally again, transport to an abattoir and slaughter for human consumption (assuming that all meat-withdrawal periods following any treatment that may have been given have expired) may be the best option for minimising financial loss.

MASTITIS

As with lameness, mastitis in the beef suckler herd does not cause a problem of the magnitude experienced in the dairy industry, perhaps because of the less prolific milk yield of beef suckler cows compared to dairy cows, perhaps because of the more frequent emptying of the udder as the calf suckles compared with the standard twice or three times a day milking of most dairy cows, and because of the irrelevance of cell counts in the beef industry.

Summer Mastitis

Dry cow or summer mastitis, however, does assume proportionally greater significance within the beef suckler herd. This is usually a consequence of a mixed infection of the mammary gland often involving *Trueperella pyogenes*, *Streptococcus dysgalactiae* and *Peptostreptococcus indolicus*, and often affects animals grazing unimproved or weed-infested pastures, with flies involved in the spread of the disease. If not identified early, thick, creamy-white pus with a characteristic smell will accumulate

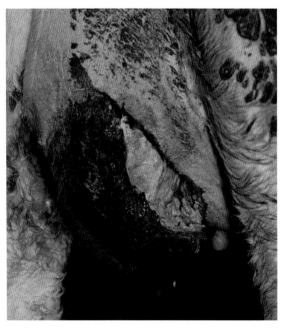

Severe mastitis that has resulted in the loss of the quarter, which is now beginning to slough.

in the affected quarter(s), which will become swollen, hot, red and painful, and the affected animal may become systemically sick.

Treatment and Prevention

Treatment of such cases is often unrewarding because of the poor penetration of antibiotics, either given into the quarter or parenterally, into the pus-filled udder, with amputation or 'splitting' of the teat required to achieve satisfactory drainage from the infected quarter. Even if the quarter is saved, milk yield from it will subsequently be reduced, which may not matter in a suckler cow if it is only one quarter that is affected, but will become significant in terms of calf performance if additional quarters are also affected. Prevention in the dairy industry has centred, for many years, on long-acting antibiotic-containing dry cow therapy and the use of teat sealants, but these are rarely used in the suckler herd. Good fly control,

however, makes sense, as does avoiding grazing high-risk pastures at high-risk times of the year.

SKIN CONDITIONS

Photosensitisation

Although primary sunburn can and does affect cattle, what is often referred to as 'sunburn' is usually photosensitisation.

Photosensitisation occurs as a consequence of the accumulation under the skin of photoactive compounds, usually porphyrins or phylloerythrins, following the increased ingestion of plants containing these compounds (primary photosensitisation), or due to liver damage reducing the ability of the organ to detoxify the compounds (secondary photosensitisation). The action of usually ultra-violet light on the photoactive compounds causes damage to the tissues in which they have been deposited, resulting

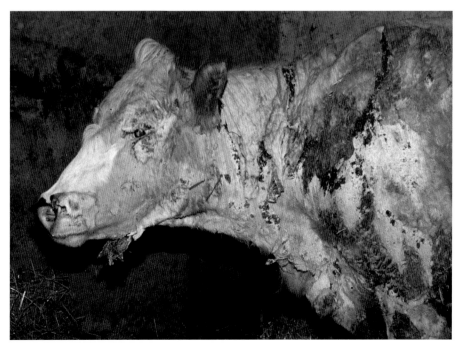

A Simmental x cow suffering severe and extensive photosensitisation. Welfare in such cases must be considered before embarking on treatment.

The cow shown in the previous image four weeks later after treatment and having been housed out of the sun.

in oedema, erythema, pruritis, and eventual sloughing of sometimes large areas of short or sparsely haired, non-pigmented skin. (More densely haired and pigmented skin usually remains unaffected because ultraviolet light cannot penetrate it to activate the photoactive compounds.)

Treatment

Treatment is symptomatic: keeping the animal(s) out of direct sunlight, applying salves to soothe the skin and prevent hardening and cracking, and controlling secondary infection and myiasis. Although healing will often exceed expectation, euthanasia should be considered on humane grounds in the most severe cases.

Warts

Warts, or cutaneous papillomas, are a consequence of infection by one of a number of papilloma viruses. Single small warts rarely cause a clinical problem (although they may, depending on the site

Extensive cutaneous warts.

involved – even small penile warts may prevent intromission or ejaculation) and are usually self-limiting, regressing in time. Accumulations of warts or larger growths, often referred to as angleberries, which most frequently occur on the ventral midline, may, however, cause problems (significant accumulations of warts on the udder and teats of breeding females may, for example, prevent calves from suckling) and may need removing. Secondary infection and myiasis, which are not uncommon complications when accumulations of warts start to spontaneously regress, may cause more significant issues than the warts themselves.

UROLITHIASIS

Urolithiasis is really a problem seen in fattening bulls rather than suckler cows and their calves, but it is included here because of the dramatic nature of the problem when it does arise. Dietary mineral imbalances in the fattening ration can result in the precipitation of a mineral sludge or 'stones' primarily composed of calcium, magnesium and phosphate around an organic core in the urine within the bladder. Although potentially an irritant to the bladder lining, clinical signs of any problem are rare. However, should one of the stones pass into the urethra it

The consequences of calculi blocking the urethra causing it to rupture, and allowing urine to escape into the surrounding tissues.

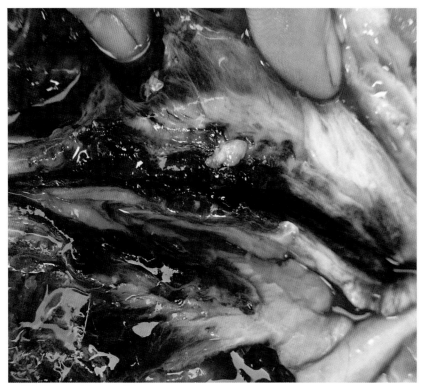

The calculus that blocked the urethra in the calf shown in the previous image, identified after euthanasia of the affected animal. Note the severe inflammation of the urethral mucosa, which gives an indication of the pain associated with this condition.

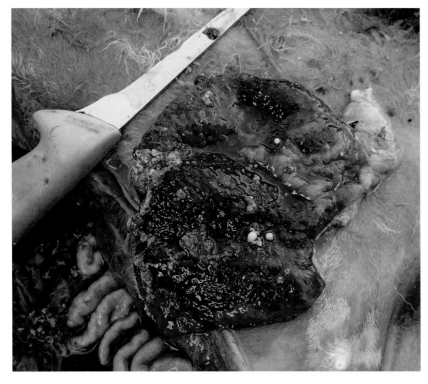

Further calculi identified within the bladder of the same animal.

A perineal urethrostomy and the placement of a catheter up the sectioned urethra and into the bladder (assuming it remains intact!) to bypass a blockage of the urethra and allow urine to be passed may be acceptable to allow resolution of the inevitable circulating uraemia; it also means that the animal may be suitable for slaughter for human consumption if it is close to finishing.

may cause a blockage; usually this occurs in an entire bull because the urethra in such animals is narrower than in castrated bulls and heifers, but whichever the case, the animal will suddenly be unable to pass urine and the situation becomes an emergency. Initial signs include abdominal discomfort and straining. The presence of calculi on dry preputial hairs may hint at the diagnosis.

Treatment and Prevention

Temporary relief may be achieved, assuming the bladder remains intact, by passing a large bore needle through the abdominal wall and into the bladder to drain the accumulated urine, while the administration of smooth muscle relaxing drugs might relax the urethra sufficiently to allow the stone to be passed. Success in such situations is not common, however, and surgery to perform a perineal urethrostomy proximal to the sigmoid flexure, the most common site at which calculi lodge in the urethra, might give a more reliable solution to salvage the animal.

In more advanced cases, where the urethra or the bladder has already ruptured, euthanasia is indicated on humane grounds.

Prevention depends on ensuring the correct ratio of calcium to phosphorus in the fattening ration (at least 1.2:1), adding salt to the ration to ensure an adequate intake of water and therefore also of urine production, and adding urinary acidifiers such as ammonium sulphate to the diet.

AFTERWORD

Much of what has been discussed above requires commitment, attention to detail and multiple handlings of both the breeding cows and their calves to accurately assess body-condition score or to measure weight, to administer prophylactic medications, vaccines or drugs to synchronise oestrus, to assess pregnancy status, or to collect samples to assess nutritional or health status. This will only be done if it can be achieved easily, is safe, and causes minimal stress to both animals and handlers alike. This requires compliant cattle, a calm manner, and a handling system and crush that allows the animals to be handled with a minimum of fuss. The importance of studying animal behaviour and then investing in a robust and well-designed handling system that utilises natural behaviours to achieve the desired outcome of easy handling to perform routine tasks aimed at achieving optimal herd performance cannot be overstated.

And when all else fails, please don't forget the vet as a source of impartial and unbiased advice. He or she may save you far more than the cost of their visit!

A well designed handling system (the first image) and a good crush (the second image) makes working with cattle much safer and easier, facilitating all aspects of herd management – and what's easy gets done (what's difficult does not!).

INDEX

First published in 2024 by
The Crowood Press Ltd
Ramsbury, Marlborough
Wiltshire SN8 2HR

enquiries@crowood.com

www.crowood.com

British Library Cataloguing-in-Publication Data
A catalogue record for this book is available from the British Library.

ISBN 978 0 7198 4393 8

Typeset by Simon and Sons

Cover design by Blue Sunflower Creative

Printed and bound in India by Thomson Press India Ltd